Integrated Packaging Systems for Transportation and Distribution

Packaging and Converting Technology

A Series of Reference Books

edited by

Harold A. Hughes
Director, School of Packaging
Michigan State University
East Lansing, Michigan

Integrated Packaging Systems for Transportation and Distribution

Charles W. Ebeling
Logistics Support Systems
Westville, Connecticut

CRC Press
Taylor & Francis Group
Boca Raton London New York

CRC Press is an imprint of the
Taylor & Francis Group, an **informa** business

CRC Press
Taylor & Francis Group
6000 Broken Sound Parkway NW, Suite 300
Boca Raton, FL 33487-2742

First issued in paperback 2019

© 1990 by Taylor & Francis Group, LLC
CRC Press is an imprint of Taylor & Francis Group, an Informa business

No claim to original U.S. Government works

ISBN-13: 978-0-8247-8343-3 (hbk)
ISBN-13: 978-0-367-40317-1 (pbk)

Library of Congress Cataloging-in-Publication Data

Ebeling, Charles W.
    Integrated packaging systems for transportation and distribution /
Charles W. Ebeling
        p.   cm. -- (Packaging and converting technology ; 3)
    Includes bibliographical references and index.
    ISBN 0-8247-8343-3
    1. Packaging. 2. Packing for shipment. 3. Packing for shipment-
-Case studies.   I. Series
TS195.2.E26   1990
658.5'64--dc20   90-36784
                                                                      CIP

Visit the Taylor & Francis Web site at
http://www.taylorandfrancis.com

and the CRC Press Web site at
http://www.crcpress.com

Preface

The purpose of this text is to provide packaging specialists and logistics managers with the knowledge and insight necessary for the practical design and sizing of bulk transport modules for effective integration into total logistical systems.

Unitized shipping, the combining of a number of smaller containers into a single, large transport module, began in the food industry in the United States in the 1950s. Handling was predominantly by lift trucks and pallets; consequently the transport modules were called palletloads.

Palletloads of food products were sized to contain enough product to gain the benefits of mechanized handling, but were small enough to provide logistical flexibility in the processing of orders and the deployment of inventory.

In the years that followed, the use of the bulk transport module concept was expanded to a wide variety of food and nonfood manufactured goods and raw materials. Bulk containers appeared that had the cubic capacity to carry the equivalent contents of an entire palletload of smaller containers of the product.

All of these bulk transport modules have one thing in common: They are too big and heavy to lift and transport by hand. They must be handled by mechanical equipment throughout the logistical process. The efficiency of the logistical system therefore is dependent upon the effective integration of the packaging and mechanical systems.

Information on the development and the evolution of the transport module concept is provided in this text in order to acquaint packaging specialists and logistics managers with the functional and economic objectives of bulk module systems. Guidelines for researching and assessing the system activities that influence the packaging design parameters for bulk modules are covered. Case histories on the development of innovative bulk module methods and systems are provided to illustrate the role of the packaging specialist in the development of the total system.

I joined General Foods Corporation in 1955 as an industrial engineer just in time to join the team that developed the first large-scale unitized shipping program in the United States. My 30-year career with the company was dedicated to the continuous research and development of new and innovative ways of handling and shipping the company's goods and materials throughout the world. During that time unitized shipping, transportation, warehousing, and materials-handling methods underwent many changes. Steadily increasing shipping volumes created a demand for the wider use of the bulk transport module concept throughout industry.

Following my retirement I founded Logistics Support Systems, a consulting service dedicated to the design and development of international bulk shipping systems. Much of the material in this book is based upon my personal experiences during my career with General Foods and my more recent projects as a consultant. A good deal of the material is taken from articles and technical papers that I have authored over the years.

My motivation for writing this book came from Mr. Edmund

A. Leonard, formerly Packaging Center Manager and Principal Scientist at General Foods Corporation. Mr. Leonard, an author of five books on packaging himself, and Adjunct Professor of Food Science at Cornell University, felt that a book that focused on integrated packaging design for unitized shipping would provide needed reference material for packaging professionals and logistics managers.

Mr. Leonard's encouragement and his help and guidance in reading and critiquing this new book on integrated packaging systems for transportation and distribution is gratefully acknowledged.

I wish to acknowledge the comments and suggestions of those individuals on the prepublication review panel who were especially helpful; their guidance was invaluable to the final editing. The distinguished panel included: Mr. Julius B. Kupersmit, Founder and President of Containair Systems Corporation of New York (Mr. Kupersmit holds over 60 patents on bulk container systems and heads a worldwide network of licensees and distributors of his unique bulk containers); James B. Holzbach, Vice President, Logistics, Kraft/General Foods, at White Plains, N.Y.; Mr. Richard Akagi, Division Manager, Packaging, American Management Association, New York, N.Y.; Mr. David K. Spencer, formerly Palletless Systems Product Development Manager, and now Regional Manager, Cascade Corporation, Chicago, Ill.; Dr. Diana Twede, Assistant Professor, School of Packaging, Michigan State University.

I am also indebted to many others who contributed ideas and materials for this book, especially Mr. James Wampler, Vice President of Basiloid Corporation, Elnora, Ind.; Mr. W. Duncan Godshall, Packaging Engineering Consultant of Madison, Wis.; and Mr. Julius Minder, a long-time friend and an international logistics systems consultant in Zurich, Switzerland.

Charles W. Ebeling

Handling barreled cement on Pier 14, the Howard Street Wharf, in San Francisco in the early twentieth century. The sailing ship in the background had just arrived from Europe with a cargo of manufactured goods and cement following a long journey around Cape Horn. The workers in the foreground are loading heavy wooden casks of cement onto a horse-drawn dray. (Photo courtesy of San Francisco Maritime National Historic Park Collection.)

Packaging in that era had to be designed to facilitate manual handling. Like cement, many goods and materials were packed in bilged wooden casks. The casks added substantial tare weight to the loads, and their round shapes did not provide the most efficient use of space in storage areas and cargo holds. They did, however, make it possible for men to move very heavy containers without the use of mechanically powered equipment.

Today most goods and materials are packaged in lightweight, square or rectangular containers and are unitized to form transport modules for mechanical handling. The functional and cost efficiencies of the handling and shipping operations are dependent on the design of the packaging and on how well the modules interface with the mechanical activities.

Contents

1

Introduction to Transport Modules

Bulk transport modules are not really new in the history of handling and shipping goods and materials. There are certain inherent benefits to packaging and shipping in bulk that have been around for centuries. Insight into the design process for bulk containers can be gained through a study of ancient shipping methods.

In the summer of 1984, U.S. and Turkish marine archaeologists began the excavation of an ancient shipwreck at Ulu Burun off the Mediterranean coast of Turkey. Artifacts and coins found on the site indicated that the ship had sunk sometime in the early 14th century B.C., thereby making it the oldest known shipwreck ever found.

Conspicuous among the scattered cargo of tin, glass and copper ingots, bronze swords and arrowheads, pieces of ivory, and pottery containers were six giant earthenware jars. Similar jars were found in other ancient wrecks along the Mediterranean and Aegean seacoasts. Scientists assumed they were used to carry fresh water for the crew.

The archaeologists at Ulu Burun maneuvered one of the big jars into a cargo net and brought it to the surface for further examination. They were astonished to find that it was packed with many smaller pieces including tableware, jugs, bowls, and other utensils. The giant jars were actually bulk shipping containers of the bronze age.

The artifacts found at Ulu Burun provided the archaeologists with new knowledge about the trade routes and cargoes of the late Bronze Age. The bulk jars open many questions in the minds of present-day packaging designers and logistics managers, such as:

> Why did the ancient shippers elect to use huge bulk containers in an age that predated modern materials-handling methods and powered equipment?

> When filled, the jars could have weighed a half ton or more. How were they moved aboard and lowered into the holds and later unloaded at their destination?

> Did the jar design serve some packaging or handling function? Why was it round instead of square or rectangular? Why did it have a wide top opening with a perimeter lip? Why was it broadest in diameter at the center and tapered to a smaller base?

> How was the top-heavy container kept upright and stable as the ship tossed and rolled on high seas?

> Finally, were there functional and economic reasons for the bulk jars that justified their use in place of many smaller containers that could be easily lifted and handled manually?

It can be assumed that clay was the only material available for the manufacture of jars. The jar shape, to a large degree, was no doubt dictated by the primitive ceramic process. The giant jar, however, could have been purposely shaped to accommodate materials handling. The round shape permitted a jar to be tipped and rolled when empty from the dock onto the vessel. The tapered shape facilitated steering it in the process. The only known materials-handling devices available at the time were wooden poles and hemp rope. The jar had a raised perimeter lip under which a loop of rope could be secured.

It can be surmised that the empty jars were rolled onto the upper deck and set upright. A loop of rope was then placed around the perimeter lip and fastened to a wooden pole. Two deck hands, one on either side of the pole, could then position the jar over a hatch and lower it into the hold below. To keep the jars upright, the tapered ends could be anchored in a bed of thorny burnet, a native shrub that is found along the Aegean Sea coastal area. Traces of the shrub were found at the wreck site, with some still packed around the jars, similar to the way paper dunnage is used to stabilize loads for shipment today.

Filling the big jars must have taken place once they were secured in position in the hold. Liquid cargo, water, or wine was probably carried aboard in smaller containers and emptied into the big jars. Smaller pottery items and other pieces of small merchandise were brought aboard by hand and packed carefully into the jars.

When the ship arrived at its destination, the cargo was very likely removed from the jars and taken ashore by similar means. Liquids were ladled into smaller containers, and other merchandise was taken out piece by piece. Loading and unloading the bulk containers as they sat in place in the hold eliminated the need for materials-handling systems to handle them loaded.

An obvious question is why did they not forego the use of large bulk jars and store the smaller vessels and containers directly in the hold? If the big jars had to be loaded and unloaded in place, the benefits of containerized bulk would seem questionable.

We can only theorize the reasons why the big bulk jars made sense. First, it may have been possible that the giant jars themselves were products to be delivered to a trade partner somewhere along the route. The cargo density in that case was increased by filling the large jars with liquid or smaller pieces of merchandise rather than waste the space inside them during the trip.

Second, assuming that was not the case and the big jars were used strictly as giant containers for the trip, what function thereby justified their use? One theory is that they made possible more effective utilization of the space available. Smaller containers found in the wreck were not designed to be stacked more than one tier high. Single tiers consequently resulted in unused space above them and reduced the amount of cargo that could be stored in the hold. Tall jars with vertical storage of smaller containers inside could have contained several times the number of smaller containers that could have been carried in single layers in the same hold. Therefore, it may be concluded that the big jars provided an ancient form of unitized shipping and stowage.

Still another function of the big jars could have been to separate merchandise into specific lots according to customer orders, types of merchandise, or destinations. Finally, the large jars, by absorbing the shocks and vibrations of the sea voyage, may have provided a means of protecting the smaller pieces from damage.

The theories concerning the handling methods and functions of the giant bulk jars may or may not be entirely accurate. It does appear, however, that many of the basic principles of integrated bulk packaging design were practiced in the 14th century B.C. Then, as now, bulk packaging must be designed as a system to interface effectively with the other functional areas of the logistical process in order:

> to accommodate materials-handling methods and equipment available,

> to permit effective utilization of cubic capacity in the transport vehicle,

> to unitize groups of items for convenience of storage and control, and

> to reduce vulnerability to product damage in transit.

It can be concluded that the durable jar containers eliminated packaging waste, since they appeared to be capable of many trips. This is evidenced by the fact that they not only protected their contents during the pounding of the storm at sea that sank the ship, but they remained intact and still protecting their cargo at the bottom of the sea for 3400 years.

If our theories are valid, the design of the ancient jars must have been done by persons familiar with the primitive materials-handling methods and the use of hemp rope and dunnage. They must have known something about the inside cubic capacity needed for the kinds of cargo transported, and they must have researched the dimensions of the holds and hatch openings aboard ship in order to size the jars correctly.

As we enter the final decade of the twentieth century A.D., the need for bulk containerized shipping is greater than ever. Over five billion people populate the world today, and over 11 billion are expected in the next century. Support of the steadily increasing population is dependent on continuous scientific and technological progress in all fields. Logistical support to speed the transport and relocation of goods and materials from supply sources to far-flung market areas is especially critical. Containerized bulk shipping can contribute much to satisfy the needs for the mass movement of goods and materials of the future.

The traditional fragmentation of the logistical process into a complex series of independent activities may discourage the initiation of system innovation or change in any one area of activity. Nevertheless, packaging is a prime element that can influence the costs and efficiencies of all other areas in the logistical process. Those who specialize in packaging design are therefore in a good position to assume the leadership needed to introduce and implement new and innovative logistical methods and systems.

Today lighter and stronger kinds of materials are available to bulk container designers. A wide variety of materials-handling methods and equipment is in use. The containers may be stacked high on single-story warehouse floors or inserted into high-rise racks by automated stacker machines (Figure 1.1). The shipping mode could be by highway, rail, ocean, air freight, and intermodal

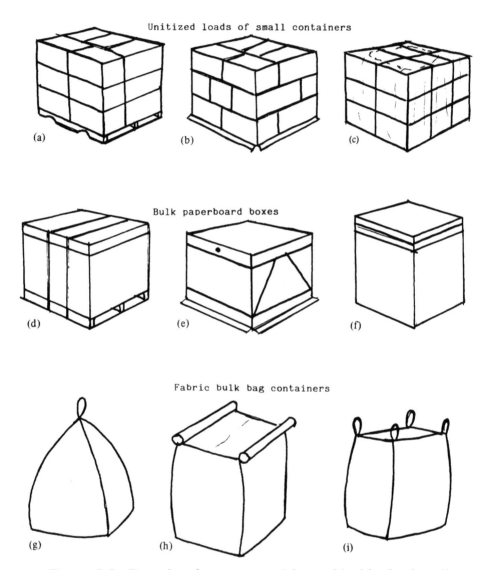

Figure 1.1 Examples of transport modules – unitized loads of small containers: (a) load on a pallet, (b) load on a slipsheet, (c) clampable load; Bulk paperboard boxes: (d) box strapped to pallet, (e) box with slipsheet base, (f) tube/cap box with top-lift fold; Fabric bulk bag containers: (g) single-loop lift strap, (h) fabric sleeves for lifting from the top, (i) four web strap loops for top lift.

combinations. Shipping distances may vary from a few miles to halfway around the world.

Today, more than ever before in history, designers of containerized bulk packaging must research and build knowledge about the entire system through which the containers they design will pass. They should have in-depth knowledge of the functional needs for packaging design in each area of activity in the logistical process in order to make recommendations on packaging that will interface well with all activities along the chain.

There are many independent activities that make up the typical logistical chain for manufactured products. The flow of goods and materials is usually set into motion by the purchasing department. The purchasing managers research the sources of supply for raw materials and supplies needed for the production or assembly process. They evaluate total procurement costs, which include the costs of packaging, storage, handling, and shipping methods. The value of the goods or materials purchased provides the purchasing manager with a guideline as to what is prudent to spend on packaging.

Packaging designers should be equipped to justify the costs of packaging. The least cost packaging that may be possible may impact handling, storage, and shipping costs unfavorably and result in high overall system costs.

Cost-efficient transportation requires effective utilization of space available inside transport vehicles, with net tonnage as close to highway weight limits as possible. The load configurations are critical and are dependent upon the proper dimensions of the unitized loads, which in turn are dependent upon the dimensions of individual containers in the unitized loads. Lightweight products must be dimensioned to fit as many products as possible into the available space. Heavier containers should be dimensioned to form efficient load configurations for uniform weight distribution throughout the vehicle. The tare weight of packaging and handling devices should be held to a minimum, since it will impact the net tonnage shipping costs. Packaging designers should have accurate information on the type of transport vehicles to be used and their inside dimensions and doorway clearances. They should be well aware of vehicle and operational load weight limitations and other restrictions.

Warehousing and storage of the goods and materials along the logistical chain will also impact costs. The packaging designer should have knowledge of the kind of warehouses and storage conditions involved. Warehousing and storage costs are based on the costs to build and maintain storage space. Efficient use of the space is dependent to a large degree on packaging design. The dimensions and stackability of unitized goods and materials are key factors in the economics of storage.

The productivity and cost efficiency of materials-handling methods that are used to load and unload transport vehicles at the docks, and to transport the goods and materials to and from storage and production locations, can also be influenced by packaging design. Bulk containers must be adaptable to mechanical handling operations for filling at supply points and for the discharge of contents at the user locations. Some containers may be vibrated during the filling process to densify the contents to achieve efficient payloads and to reduce packaging costs per ton shipped. On the discharge ends of the chain, mechanical tilt devices may be used to invert containers to allow contents to flow out by gravity. Other locations may elect to use air transfer or pumping systems.

Distribution operations for finished products are usually characterized by multiple handling, shipping, and storage activities. The durability of packaging to stand up well throughout the distribution process is critical. The adaptability of smaller containers to efficient unitload patterns is extremely important to handling and shipping efficiency. The design of the unitload for distribution packaging should be part of the total packaging design process.

Overall, packaging must be adequate for the protection and containment of goods and materials through handling, storage, shipping, and distribution operations. Too much packaging, however, will add to tare weight and become a costly environmental nuisance to dispose of at the user end of the logistical chain.

Too often, packaging specialists are not sufficiently involved in the total logistical process to recognize the impact that packag-

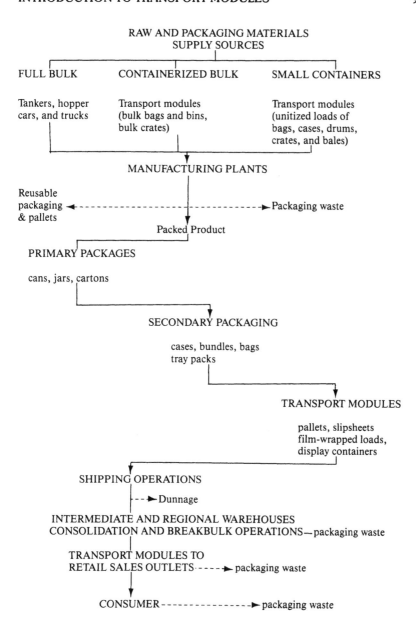

Figure 1.2 Packaging in a typical logistical system.

ing design can have on the costs of the other logistical activities. Even if involved, they may not be in a position in the organizational structure to bring about the compromises that must be made for overall system cost benefits. In any case, integrated bulk packaging system design requires the packaging deisgner to be thoroughly familiar with the total logistical system in order to recommend packaging alternatives to management.

Packaging designers should have in-depth knowledge of the different kinds of bulk containers and systems. They should be well informed on the alternative unitized methods of shipping and the different kinds of transport equipment. This kind of knowledge will put them in a good position to advise management on packaging, as well as to design transport modules that can be integrated cost effectively into the total system (Figure 1.2).

The intent of the following chapters is to provide the packaging specialist with information about the development and evolution of the different kinds of unitized and bulk container transport modules and the related handling methods and equipment.

Background and general information is provided to help the reader understand why transport modules must be designed as a total system.

2

Unitized Handling and Shipping

DEFINITION AND BACKGROUND INFORMATION

Unitized handling is one of the most significant developments in the history of shipping goods and materials. Stacking a number of objects on a tray for transport from one room to another is a simple form of unitized handling. The larger the tray, the greater the number of objects that can be carried. There are of course limits due to the size and weight of the objects.

Industrial unitized handling is the combining of a number of smaller containers into one large module for mechanical handling by lift trucks or other kinds of materials-handling equipment. The first carrying bases for unitized loads were low platforms known as skids. They consisted of several wide boards nailed across two runner boards. A manually powered hydraulic jack mounted on wheels would be pushed between the runners under the top boards.

The skid with its load would then be jacked up over the wheels to enable it to be pushed or pulled.

The urgencies of World War II gave impetus to the development of unitized methods. The simple hydraulic skid jacks evolved into fork-lift trucks, making it possible to raise heavy unitized loads and stack one on top of the other. Skids evolved into pallets as crossboards were nailed across the bottom of the runners to provide for the stability of top loads in stacks of two or more.

In the decade that followed the end of World War II, fork trucks and pallets came into widespread use. Since this predated shipping on pallets in the food industry, the pallet dimensions depended upon local preference. Length and width dimensions ranged from 48×96 in. (1219×2438 mm) for large stevedore pallets at the docks to 36×36 in. (914×914 mm) for smaller warehouses handling large numbers of products that required a relatively large number of warehouse aisle facings. The size of the pallet was dictated by the economics of mechanical handling. If too large, the fork-lift trucks that were required to handle them would be too big and cumbersome for operation inside warehouses. On the other hand, if the pallets were too small, the number of lift trucks and operators required could offset the economic benefits of mechanical handling.

Unitized shipping on pallets did not make inroads into industrial shipping until the late 1950s. It began with the adaptation of wooden pallets for the unitized shipping of cased packaged goods in the food industry. Palletized shipping offered the food industry a much improved alternative to the manual loading and unloading of transport vehicles. It had, however, many shortcomings, which eventually led to a search for unitized methods that did not require the use of pallets.

The palletized method and its limitations as a unitized shipping system can best be understood if we review its progress from its beginning to the 1980s.

UNITIZED SHIPPING ON PALLETS

Food industry distribution methods in the 1950s were labor intensive and slow. Then, as now, the primary or retail packages were packed mostly 12 to 24 per shipping container. The containers were usually hand piled onto pallets at the ends of production lines, and the loaded pallets were taken by fork-lift trucks to warehouse holding areas prior to shipment. Fork-lift trucks and hand-operated pallet carriers were used to shuttle the products from the warehouses to the shipping docks, where crews of warehouse workers would manually transfer the shipping containers into rail cars or highway trailers. Empty pallets were then accumulated and taken back to the production lines for reuse. When the transport vehicles arrived at regional warehouses, the shipping cases were manually transferred back onto pallets for warehouse storage.

The process had to be repeated again as consolidated shipments moved from manufacturers' regional warehouses to distributor terminals where pallet loads were broken down and products assembled into retail order lots for delivery to supermarkets and other retail sales outlets.

As production line volumes increased with automated and new, faster packaging machinery, and greater volumes of goods were going through the distribution channels, it became apparent that the shipping docks were becoming the bottlenecks of the industry. Not only were labor costs to hand load and unload transport vehicles high, but the time involved tied up transport vehicles at the docks, sometimes for long periods of time. That limited the amount of revenue a vehicle could produce and impacted transportation rates unfavorably. Furthermore, long delays getting the products across shipping docks dictated the need for carrying increased inventory to satisfy sales demands.

Soon the question was asked as to why the cases could not be left on the pallets, loaded by fork-lift truck, and shipped palletized. Studies by industrial engineers (Figure 2.1) confirmed that shipping on pallets could eliminate bottlenecks on the shipping docks and increase productivity five to 10 times. The findings excited management's interest and resources were provided to develop and implement a food-industry palletized shipping program.

Conversion to palletized shipping was no simple matter. It took several years before a viable program would emerge. An early obstacle was getting the food industry to agree on a standard pallet.

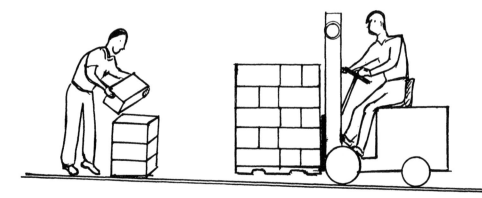

MANUAL LOADING
8 pallets, 480 cases per
man hour

MECHANICAL LOADING
50 pallets, 3000 cases
per man hour

AVERAGE CASE SIZE AND WEIGHT

Size case – 18×12×7.5 in. (457×305×190 mm)
Case weight – 22 lbs (9.97 kg)

TRANSPORT MODULE

– GF wooden pallet
        48×40×5.5 in. (1219×1016×140 mm)
        tare weight – approximately 80 lbs (36.3 kg)

– Load pattern
        # cases/tier –   10
        # tier/pallet –   6
        # cases/pallet – 60

– Load density
        net weight/palletload – 1320 lbs (599 kg)
        gross weight/palletload – 1400 lbs (635 kg)

– Shipping method
        DFB railcar – carloads
        pallets two high in stacks, 28 stacks,
        56 pallets/carload

– Storage method
        pallets stacked up to four high in warehouses.

The original GF standard 48×40 in. pallet became the industry standard later known as the GMA (Grocery Manufacturers Association) pallet.

The wide variety of case sizes and shapes resulted in few perfect module dimensions. The objective for pallet patterns was to come as closely as possible to the standard 48×40 in. (1219×1016 mm) base dimensions.

Figure 2.1   Food industry palletized shipping program – 1962.

14

Just about every manufacturing plant and warehouse had its own pallet type and size to suit its particular line of products. Decisions made at that time still impact the way goods and materials are unitized and shipped today.

Based on the interior dimensions of rail cars and truck trailers of the time, agreement was finally reached on a pallet dimensioned 48×40 in. (1219×1016 mm). Two 48 in. lengths could be placed adjacent across the inside width of a standard rail car, and two pallets with 40 in. widths adjacent would fit into typical highway trailers. A combination of one 40 in. side adjacent to a 48 in. side, in a pinwheel configuration, made it possible to increase the payloads by the addition of one or more pallet positions.

Decisions and compromises had to be made on the design and type of pallet. The pallet had to be capable of stacking unitized loads weighing from a few hundred pounds to a ton at least four high to ensure efficient use of warehouse space. The amount of board surface is critical to distribute the pressure of the unitload evenly over the pallet and to prevent damage to shipping cases. The pallet must be designed to allow the entry of lift-truck tines from four sides and the entry of the wheeled arm extensions of low-lift-type pallet carriers (Figures 2.2 and 2.3) from either end.

Many different kinds and designs of pallets were tested over several years before the final decision was made on what became the standard grocery pallet. Research to find a low-cost pallet that could be discarded after one trip proved to be fruitless. Cheap pallets could not take the abuse of transit. They broke up too easily in transit, causing damage to products. It was eventually recognized that a pallet designed to meet the functional requirements of the grocery products industry would cost substantially more than anticipated.

To ensure that the pallet would be durable enough, the use of hardwoods was mandatory. Following hundreds of test shipments, a final list of detailed specifications covering the dimensions of wooden pieces, types of nails, and recommended manufacturing and assembly procedures was firmed up. The cost of such pallets dictated that they be recovered after each trip and reused several times to be affordable.

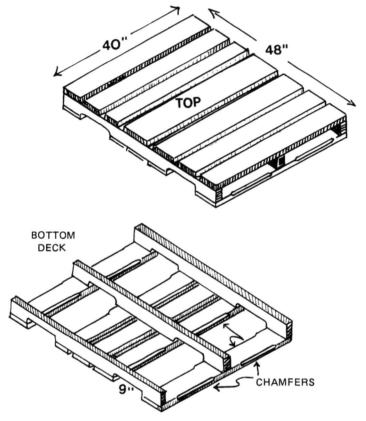

40"   48"   TOP

BOTTOM
DECK

9"   CHAMFERS

Figure 2.2

That necessitated the establishment of an industrywide pallet exchange system. Shippers and receivers agreed to jointly support the exchange program. Empty pallets of the same quality and specifications as those received under loads would be returned in the transport vehicles to the shipping source.

It was recognized during test shipments of palletized loads that existing rail and truck transport equipment did not lend itself to efficient palletized shipping. Narrow doorways of boxcars made the maneuvering of lift trucks entering with pallet loads difficult. Flooring in most rail cars and highway trailers was too light to support the weight of lift trucks carrying full palletloads. During

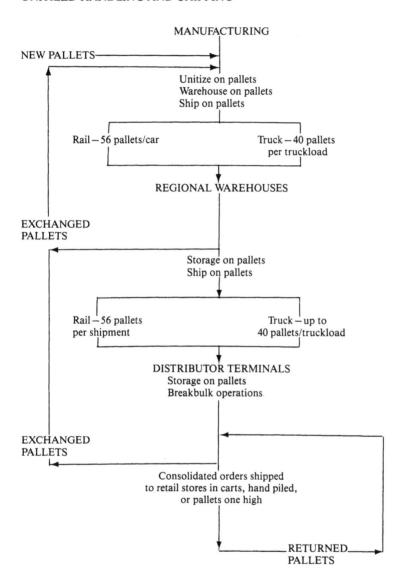

Figure 2.3   Pallet exchange system (food industry) — 1960s.

the 1960s thousands of new rail cars, specially designed for the transport of unitized loads on pallets, were put into service. Special features included heavy flooring, bulkheads that could be locked into place to restrain palletloads into three compartments, extendable wall boards to close the gap between pallets and the sides of the cars, and special suspension systems to cushion the ride.

Investment in the cars was justified by the increased revenue produced by the palletized shipping program. Cars containing 56 palletloads of products could be loaded or unloaded by fork-lift trucks in less than 2 hours, thereby freeing them up for continuous productive use. Railroad tariffs were written to allow 56 standard pallets to ride as part of the vehicle, which permitted shippers to return empty pallets at no charge.

Special highway vehicles were acquired to transport palletloads on shorter hauls to distributor terminals. These vehicles had doorway openings 102 in. (2590 mm) high to allow a lift truck carrying a stack of two pallets to enter. Floor decking was designed to bear the weight of a standard lift truck and two pallet stacks of the heaviest product.

Most new warehouses built during the time were designed to accommodate palletized storage and shipping. Floor levels were raised to the height of the decks of truck trailers, generally 48–54 in. (1219–1372 mm). Rail tracks were brought inside many new warehouses, and the shipping docks were raised 42 in. (1067 mm) above the rails.

The structural columns that supported the roofs over the single-story structures were spaced to accommodate the storage of products in 1000 sq ft (92.9 m²) storage bays. Typical bays were 40 ft wide and 25 ft deep (12.2×7.62 m) to accommodate 60 stack storage blocks. Ceilings permitted the storage of products in stacks 18.5–21 ft high (5.64×6.4 m) in most warehouses. Steel racks, used most commonly in distributor terminals, were designed for the storage of 48×40 in. (1219×1016 mm) standard palletload modules.

Changes in materials-handling equipment were necessary to adapt to palletized handling and shipping. Fork-lift trucks were equipped with side shifters. These are devices that can shift loads on forks laterally, 4 in. (102 mm) side to side, to facilitate maneuvering in tight areas. Multistage masts were installed on fork-lift trucks to enable them to extend upward to high stack palletloads in warehouses, and then collapse low enough to permit the lift trucks to enter the doorways of rail cars or highway vehicles.

The potential health hazard involved in the operation of gas-powered lift trucks in areas of limited ventilation during the loading and unloading of vehicles led to the development and installation of solid-state, high-performance, battery-powered lift trucks in many warehouses in the 1960s.

Standardization on the 48×40 in. pallet speeded the installation of pallet loading machines at manufacturing plants. The tedious, backbreaking work associated with hand loading cases onto pallets was eliminated on just about all high-volume production lines. Cased products were conveyed in a line from the packing lines directly into machines that automatically arranged the cases in proper configuration, and tier by tier on the pallets.

By the early 1960s palletized shipping had been accepted by most large manufacturers and distributors in the food industry. Palletized shipping eliminated bottlenecks on the shipping docks, reduced handling costs, and improved the productivity of transport vehicles. Still, it was not without serious problems.

The packaging and the design of the shipping cases was, and is to this day, a key problem for palletized shipping. Few shipping cases, of the several thousand different size cases in the packaged goods distribution systems, adapt to a reasonable pattern to fit precisely onto the 48×40 in. pallet surface. The outside dimensions of the shipping cases are dictated by the primary package size, the number in the pack, and the pack configuration.

Shippers try to form the cases into load patterns compatible with the 48×40 in. module, with as little as possible overhanging the pallet sides or underhanging the pallet surface. Overhanging loads creases the cases in the bottom tiers, weakens the load, and causes stacks of two or more high to lean. Underhanging allows the cases to shift about in transit, which can also cause damage to cases and result in unstable stacks.

The stability of unitloads can be increased if rectangular shipping cases are interlocked by reversing the direction of each tier in
the load and securing the top tier with a perimeter tape. The
interlocked pattern holds the load together in transit, however, the
degree of efficiency depends upon the weight of the case and the
amount of overlap between the cases in the pattern. The lighter the
case, the greater the amount of overlapping that is required. Lockload patterns, unfortunately, can make the lower cases in the load
vulnerable to damage, particularly with very heavy cases. Divider
sheets of solid fiber or corrugated board can be used to separate
the tiers and prevent the weakening of the top flaps of cases that
are directly under the overlapping edges of cases in the tier above.
That, of course, adds to the costs of shipping, but may prevent
substantial damage.

The building of modules without reversing every other tier
requires the cases to be secured in place firmly. This method is
commonly called column stacking since the cases are arranged in a
series of columns with each case directly on top the other. Load
pressures are therefore transferred down through the walls and
corners of the containers, and top-to-bottom compression resistance is substantially increased in comparison with lockload patterns. Column stacking became especially effective when shrink-
and-stretch film overwrapping became available.

The dimensions of many cases, however, do not form well
into column stacks that are within the tolerances of standard shipping module dimensions. Consequently, lockload patterns are still
in use today. Some lockload patterns have void spaces within the
load. This should be avoided wherever possible, since the void
spaces do not provide a good bearing surface for the top loads,
and it is possible that the containers may shift about in transit and
cause damage. The use of stretch-film overwrapping will help stabilize the interlocked tiers in such patterns and will reduce the
amount of shifting about that the containers are subjected to in
transit.

Package specialists should develop the best module pattern
possible for the containers they design and also put the modules

through a series of dynamic load tests to ensure that they have the best combination possible. The securing of the load by film overwrapping should be included in the packaging specifications.

In order to arrange certain size cases into an interlocked pattern, void space must sometimes be left between some cases throughout the load. This, too, can result in damage and will limit the net payloads handled and shipped.

The first of many computerized pallet programs appeared in the mid 1960s. Outside dimensions of the shipping cases were fed into a computer, which would search all possibilities for patterns to fit the standard 48×40 in. module. Adequate tolerances for overhang and underhang had to be written into the computer program, otherwise the computer could seldom find a pattern. There were instances in which a very small change in package size could have resulted in a more efficient pallet pattern. Most product managers, however, were reluctant to change the package sizes already established, since even small changes of dimensions could change the appearance of the store shelf displays.

Later on, computer programs were developed to assist packaging designers to analyze alternative dimensions of new retail packages and to select the case pack configurations that provide case sizes that adapt well to the standard palletload patterns. Such tools have helped but have not eliminated all of the pallet pattern-design problems.

Lightweight, bulky shipping cases were another problem for palletized shipping. The space that the pallets took up under the loads limited the number of such cases that could be shipped in the space available. Many shippers elected to continue hand loading and unloading these kinds of products to avoid transportation penalties.

As the number of shippers in the palletized program expanded, and hundreds of shipping locations participated in the pallet exchange system, control of the pallet specifications became very difficult. Softwood boards and wide spaces between the surface boards appeared in the system. Not only did the lack of bearing surface cause damage to many products, but the cheaper pallets did not meet the life expectancy of the pallets with proper specifications. Damaged pallets with broken boards and protruding nails were common in the system.

An industry committee was subsequently established in an effort to enforce the use of the original specifications. It found that compromises on the use of hardwoods were necessary due to shortages of specified hardwoods. Up to 18% of all the lumber produced in the United States was going into pallet production, and half the board feet of hardwood lumber each year was being consumed by the pallet industry. The continued demand for pallets complicated getting and maintaining a flow of good pallets into the system. Not only were pallet repair costs increasing, but many shippers simply could not take pallets out of service long enough for proper repairs. The committee eventually concluded that the original pallet standards could not be effectively maintained in the industrywide exchange system, and the committee was eventually disbanded. Shippers who had been putting costlier pallets into the program decided they could only contribute the kinds of pallets they received in exchange. This led to the further deterioration of the quality of pallets in the exchange system.

Along with the deterioration of the pallet system, sanitation managers of many food manufacturers were becoming concerned with the condition of wooden pallets used under food products. Piles of empty pallets were often stacked in the yards outside the plants and warehouses due to fire insurance requirements. There appeared to be no cost-effective way to clean and sterilize wooden pallets, which often were subjected to contamination from rodents and bird droppings. Pieces of broken boards, splinters, and loose nails were commonplace around storage areas.

At attempt was made to get a suitable plastic pallet developed to replace wooden pallets in the food industry system. It was thought that plastic pallets would be more durable than wooden pallets, and more sanitary for use around food products since they could be steam cleaned.

A number of plastic pallets were eventually tested, but none lived up to expectations. The plastic materials gradually deformed under heavy loads, both in floor stacks and in racks. It was found, too, that plastic pallets were nearly as vulnerable to fork truck damage as wooden pallets. Even without the functional shortcomings, it was questionable if shippers who were putting the cheapest wooden pallets they could buy into the exchange system would consider exchanging plastic pallets, which required an investment several times that of wooden pallets.

Plastic pallets eventually gained limited use inside some plants and warehouses, but were never seriously considered as an alternative to wooden pallets in the exchange system.

In the early 1970s it was recognized by most shippers that something must be done about wooden pallets. Wooden pallets had provided a simple and easy-to-handle base for unitized loads in manufacturing plants and warehouses during the period of industrial expansion following World War II. Their troubles began when they were taken out of the nation's plants and warehouses in the 1960s to transport unitized loads of products over rail and highway. Many large manufacturers began the search for a viable unitized system that would not require the use of wooden pallets under the loads.

3

Alternative Unitized Systems

As the pallet exchange program in the United States expanded to include hundreds of shipping locations, control over the quality and specifications of the pallets in the system became increasingly difficult. Despite the problems, unitized shipping and handling had become indispensable to the cost-efficient distribution of packaged goods. A return to the slow, labor-intensive hand methods of loading and unloading was unthinkable. New ideas for unitized shipping methods that did not require pallets under the loads appeared continually. A few of the concepts are worth mentioning because they provide insight into the difficulties involved in developing unitized systems that are as universally applicable as the pallet method.

A system developed by the Unit Load Car Corporation of Chicago introduced the idea of palletless modules riding on special decking in rail cars and trucks that would eliminate the need for

any kind of base under the loads. It consisted of 48×40 in. (122×
102 cm) high-tensile steel channelled platforms that were placed
under the modules for transport inside the plants and warehouses.
Matching channels were permanently installed in the transport
vehicles. The platforms consisted of nine steel tubes, each 48 in.
(1219 mm) long and 2.375 in. (60 mm) square (Figure 3.1).

The tubes were welded onto a thin steel sheet and were spaced
2.375 in. (60 mm) apart. Eight narrow fork-lift truck tines could
be inserted into the tubes to carry the platform under the load or
into the open channels between the tubes so that the load could be
lifted free of the platform. This technique became known as the
"take-it-or-leave-it" method. The steel platforms were used in the
storage areas and for the transport of modules between storage
areas and shipping docks. On the shipping docks the loads were
removed by the take-it-or-leave-it method and loaded onto the
channelled flooring of the rail cars or trailers. At receiving docks
the modules were accessed again by narrow tines and removed to
channelled platforms on the dock for transfer to the storage areas.

A number of shipping tests were made with prototype equipment in the 1960s. The concept was eventually rejected as an alternative to wooden pallets for several reasons, which included:

> Small cased goods did not adapt well to handling on the
> narrow channels. Overhanging ends could drop into chan-
> nels and block the entry of the lift-truck tines. Weak cases
> could deform into the open channels and also block the
> entry of tines, or make withdrawal of them from under the
> loads difficult.

> Tall modules were the exception. Most palletloads were less
> than 48 in. (122 cm) tall and consequently were shipped
> two high in stacks in the transport vehicles. In order to
> separate two modules in a stack it would have been neces-
> sary to use a mobile platform under the top load. That
> defeated the prime objective of the system, which was to
> eliminate the use of any carrying base under the modules.
> The possibility of installing double decking with separate

(a)

(b)

Figure 3.1 Multi Tine, "take-it-or-leave-it" system — 1960.
Trade photos.

channels on each deck in the rail cars and trailers was considered, but it was entirely impractical due to cost and added tare weight.

It would have required an enormous investment to build an infrastructure of steel-channelled rail cars and trucks, channelled platforms, and racking in warehouses.

The long, narrow tines of the lift-truck attachments were difficult to maintain in good alignment.

The steel channels added considerably more weight to the transportation equipment compared with the tare weight of wooden pallets. This could have reduced net payloads in highway vehicles due to the over-road weight restrictions.

Some workers at a test location felt the channels were a safety hazard, since workers walking over the channels could trip or twist their ankles.

Among other ideas for palletless unitization at the time was a method introduced by the Hyster Company that involved the encapsulation of the entire unitload in a sealed plastic bag. A vacuum pumping system was then used to draw a vacuum through a series of small holes in the bag to bind the load by air pressure onto a carrying plate mounted on the lift truck. The method proved to be impractical since the power required to operate the on-board vacuum pump was costly and the time it took to draw the vacuum was time consuming. Also, fragile products could be too easily crushed and damaged as the powerful vacuum was drawn.

Still another method that came out of the 1950s and 1960s, with which packaging and logistics people should be familiar, is the skee pallet or skee sheet. While the skee method did not revolutionize unitized shipping, it filled a need that existed at the time for an alternative to pallets and slipsheets. The method is still in use today and new applications continue to appear each year.

A skee sheet is a base for a transport module. It appears at first glance to be a paperboard slipsheet. It is made from paper overlaid veneer, which consists of a center core of rotary-cut hardwood veneer to which a layer of kraft paper is laminated to each side. Molded rigid side tabs on one or more sides curve up around the base of a transport module to protect it as tapered fork-lift tines chisel under it to lift the load (Figure 3.2).

The skee sheet began as a new kind of industrial packaging material in the 1950s and was originally used as the panel material in cleated wooden crates. The crates were used extensively for overseas military shipments during the Korean war. In the mid 1950s machinery was developed to bend and mold flat sections of the material into U-shaped channels, into which a variety of hardware items such as moldings and metal weather stripping were packaged for shipment. Some of the boxes were as long as 22 ft. and carried loads up to 300 lb.

Its adaptation to use as a materials-handling device occurred in 1956 at a chemical company plant in New Orleans. The plant had converted from pallets to paper slipsheets for bagged materials. At that time neither slipsheets nor slipsheet-handling equipment had been perfected. The handling attachments were costly and difficult to operate and maintain. Neither plastic nor moisture-proof slipsheets were yet available, and the tabs

Figure 3.2 Photos courtesy Elberta Crate & Box Company, Dundee, Illinois.

of the slipsheets under loads in the hot and humid storage areas of the plant absorbed moisture and ripped off when gripped and pulled.

The purchasing manager of the company heard about the wood veneer sheets and decided to experiment with their use as a moisture-proof base in place of paper slipsheets. Unlike slipsheets, the end tabs could not be gripped and pulled, but it was found that tapered fork tines could chisel underneath the loads without causing damage to the products. It appeared that many of the benefits of slipsheet handling could be possible without the investment in costly special handling attachments.

The prototype sheets had molded curved tabs on one side that bore a resemblance to skis, and the workers at the New Orleans plant began calling them ski pallets. The Elberta Crate and Box people decided to register the word *ski* as a trade name for the sheets, but it was not available. They consequently changed the spelling to *skee*, and it became their registered trademark.

Paper slipsheets and their handling equipment were improved vastly over the years, but skee sheets had found their niche as a cost-effective alternative for certain products and for certain shipping conditions. A variety of goods and materials are shipped on skee sheets today, including bagged chemicals, cases of light bulbs, cartons of fluorescent tubes, and printed magazines. Skee applications have been growing for shipments of products in ISO intermodal containers for international shipments. They eliminate the hand stacking of products, and the handling equipment is relatively simple (Figure 3.3).

Skee sheets are manufactured at the Brainbridge, Ga. plant of Elberta Crate & Box Company. They are marketed through a sales office in Dundee, Ill. and an affiliate company, Skee Pallet Japan.

The packaging requirements that are essential for trouble-free skee handling are the same as for slipsheet handling. The transport module must be secured to the skee sheet with banding, stretch-film overwrap, or adhesives.

The packaging specialist should also be familiar with the operational requirements for skee sheets. These include:

Figure 3.3   Handling a load of bags on a skee sheet.

Accessing loads: The lift truck must be equipped with polished, tapered-tip fork tines or platens. The weight of the module handled should be no more than 3000 lb (1361 kg). The load to be accessed must be positioned against a backstop such as a wall or another load. If the combined weight of two loads in a stack exceeds 3000 lb, then each must be handled separately. A separator skee sheet with the tab pointed down should be placed on the top of the bottom load.

On the shipping end, the modules can be staged on take-it-or-leave-it pallets at the dock, in which case a backstop is not required. However, separator sheets between heavy modules in stacks of two must still be used to facilitate unloading one module at a time on the receiving end.

Depositing loads: When chiselling under a load, the lift-truck operator should allow the load to overhang the tips of the fork tines 6–8 in. (15–20 cm). When the load is deposited, it is first tilted forward to contact the flooring at the front end in order to build friction to hold the load in place as the tines are backed out. Sometimes a gap is left between the loads in a line as this is done, but it can be closed by pushing the load forward when the succeeding load is deposited.

Separation of loads in a stack: Narrow tines can sometimes get out of alignment and can make entry of the tips under the load difficult. A technique used in that case is to approach the load at an angle, allowing one tine to penetrate at a time, and then raising or lowering the other tine, as may be necessary, as the lift truck is driven forward.

The person responsible for the installation of a skee system should be aware of the pitfalls. The relative simplicity of the operational technique may appear to lessen the need for operator training and follow-up in order to ensure that proper equipment is being used. It should never be taken for granted that all shipping dock personnel are familiar with the skee method, or that they will in all cases figure out how to use it properly.

An excellent example of the problems that can be experienced took place at a New Jersey distribution center during a test shipment of heavy products on skee sheets in 1985. The transport modules weighed a ton each and were normally double stacked on cheap wooden pallets for loading into ocean containers for dispatch to a Middle East location. The pallets made loading easy, but the space they took up under the modules reduced the overall net payload, which added to shipping costs. Also, many of the pallets broke up during transit, and there were continuous complaints of substantial product damage from the receiver. Slipsheets were suggested, but the receiver did not have slipsheet-handling equipment on his dock and would not equip his lift trucks with the necessary attachments for the limited volume received.

Skee sheets seemed to be the ideal solution in this case. They were a much less costly alternative to the use of more durable but costly wooden pallets or to the hand-stacking alternative to pallets.

A few skee sheets were purchased for a test shipment and sent to the shipping dock with instructions to transfer loads onto the sheets and to scoop under them with the lift-truck forks in order to carry and deposit them into the ocean container. The night before the shipping date, the loads were transferred from pallets onto the sheets with the use of a carton clamp attachment. They were stacked two high against a wall along the dock.

The next morning a lift-truck operator, who had received only sketchy instructions on handling the loads, attempted to chisel underneath the first stack with his lift-truck tines. The tips of the tines on his truck were, unfortunately, not the tapered kind, but instead had thick, blunt ends. His attempts to force them under the stack were fruitless because the ends of the tines jammed against the front tab of the bottom load. The operator was determined to somehow get the tines under the double-stack load. Therefore, he backed up his truck several feet, aimed the fork tines at the bottom of the stack and came forward at full speed. The blunt tips of the tines crashed into the skee tab and crumbled it under the load, along with bits of the cases and some of the product from the bottom tier.

Eventually, properly tapered and polished fork tines were acquired and some operating techniques were discussed with the lift-truck operator. The tapered tines easily penetrated under the loads, and the container was loaded with little difficulty.

ECONOMICS OF SKEE SHEETS

Skee sheets are currently priced in a range that is competitive with very heavy-duty paperboard slipsheets. There are certain functional advantages that favor the slipsheet method over the skee sheets. The volume shipped is usually a key determining factor in the method selected if both are being considered as alternatives to pallets or hand-stacking methods. Where volume does not support the installation of special slipsheet-handling equipment, the skee-sheet method may offer a more cost-effective system alternative to palletized shipping.

The carton clamp and the slipsheet palletless methods of handling and shipping were introduced about the same time as the standard 48×40 in. grocery industry pallet in the United States. These two methods have passed the test of time and have come into widespread use over the years.

Packaging specialists who design transport modules should be well acquainted with clamp- and slipsheet-handling equipment and operating techniques. It is especially important that they understand how the packaging they design can impact the related handling efficiencies of the methods. The clamp and the slipsheet methods are, therefore, described in depth in the following two chapters.

4

Unitized Handling by Clamp

The clamp-handling concept can be best understood with a simple demonstration. Position four small cartons, side by side, in a line, on a table. Place your hands flat against the sides of the outer two. Keep your hands flat while pressing steadily inward until all four cartons can be lifted without slippage. The heavier the cartons the greater the pressure needed, since the pressure applied by your hands transfers through the outside cartons to create the friction needed to hold the two in the middle in place.

Clamp attachments for lift trucks were originally designed to clamp cotton bales and large heavy rolls of paper. These attachments consist of two metal plates lined with rubber pads on the inside surfaces. These pads are placed vertically around the load to be lifted and then hydraulically compressed together to create enough friction to hold the load in place as it is lifted. The proper amount of pressure is critical, however, since too little pressure will allow the load to slip free during lifting, and too much pressure could crease or damage it (Figure 4.1).

With the emergence of unitized shipping, and the problems with pallets, some shippers experimented with the use of clamps for handling unitloads of shipping containers. Packaging design and dimensions were critical to clamp handling.

> Cases in a standard 48×40 in. module had to be large enough to create sufficient friction to hold the cases in a load together during clamp compression. Generally, cases less than 5 in. (127 mm) in height could not be clamped.

Unitloads with voids in the load patterns, and patterns that

Figure 4.1 A clamp truck. Photo courtesy Cascade Corporation, Portland, Oregon.

did not have uniform surfaces on the sides, could not be handled by the clamp method.

Containers within the load had to have sufficient compression resistance in their walls to withstand the pressure of the clamps without creasing or damaging the product.

Large, bulky, lightweight shipping cases, which could be formed into unitloads with even surfaces on at least two sides, were easily handled by clamp. Smaller, heavier cases required greater clamp pressure to build enough friction between the cases in the load. Three-way, pressure-relief valves were used in some operations to regulate the pressure required for a range of light to heavy clamp loads. This helped avoid the use of too much clamp pressure, known as overclamping, which can result in product damage. Other users settled on a simpler fixed-pressure relief setting for their entire range of products. Lift-truck operators were trained to apply just the right amount of clamp pressure needed for specific products.

The clamping of cased goods was introduced as an internal warehousing system for the palletless storage of large, bulky shipping cases. The clamped products were stacked by clamps on warehouse floors and transferred onto pallets for outbound shipments. The elimination of pallets under this kind of product in the warehouses reduced the number of pallets needed for the total system.

Originally there was no intent to load and unload transport vehicles by clamp, nor to attempt to develop clamping into an alternative unitized method to replace the use of wooden pallets for shipping operations.

With the failure of the pallet exchange system and the desperate need to develop alternative unitized shipping methods, the question was asked whether the method could be applied to a wider range of products and adapted to shipping as well as to storage operations. Experimentation with shrink-film overwrapping at General Foods Corporation in the early 1970s contributed a great deal to the advancement of clamp methods. About 75% of the company's line of products could not be formed into clampable unitloads. Many products were shipped in very small containers that did not have sufficient end surfaces to create an adequate frictional bond in a clamp-load configuration. A product especially difficult to clamp were small packages of the company's Jell-o dessert line, which were wrapped in kraft paper bundles. It was realized that if unitized loads of such bundles could be made clampable, then just about any product in the entire line of products could be made clampable.

A method was subsequently developed that essentially turned the unitload of small containers into one large container. A sheet of paperboard was placed under the load and the load encapsulated in shrink film.

The method was demonstrated using a revolving clamp fork attachment on a lift truck. The paper sheet was placed on top of a palletload of the product. A plastic polyethylene bag was then fitted over the paper sheet and pulled down over the entire load. The load was then positioned onto a conveyor and put through a heat tunnel to shrink the film tightly around it. As it emerged from the tunnel the revolving clamp fork attachment was used to turn it upside down so the pallet could be manually removed from the top.

The method worked, and although the development of special equipment was required, it appeared to be affordable in comparison with the costs of pallets. Before going ahead with the further development of a clamp shipping program, however, some fire hazard concerns about film-wrapped unitized storage had to be addressed. The key concern was based on the theory that fires are brought under control in storage areas by the saturation of shipping cases with water released from sprinkler systems.

The plastic film wrap on products around the fire area would shed the water and keep the products dry. When the fire reached and burned off the film, the dry products inside would make control of the fire very difficult.

In 1971, a hazards analysis test program was carried out at the Factory Mutual Fire Research Center at West Glocester, Rhode Island. The test objective was to provide an evaluation of the potential hazards presented by film-wrapped unitload stacks on warehouse floors and to assess the relative capability of existing fire protection systems to cope with the fire hazards that might become evident.

The large-scale tests involved the use of pallet-size metal bins wrapped in corrugated material to simulate unitloads of products. The bins were stacked in a typical warehouse-storage block configuration, four per stack, three rows of stacks across and six deep. A single stack was positioned 11 ft from the main storage block to simulate an aisle and to test the adequacy of aisle width for a fire break.

Two large-scale tests were made. The first involved the pallet-less stacks in which each module was wrapped in shrink film. The second test, a control test, involved the same type of modules unwrapped and on standard wooden pallets. In the tests, a fire was ignited between two rows near the center of the storage block. Ignition intensity was the same for each test.

As expected, the fire in the film-wrapped loads shot up the flue rapidly as the film was consumed. The sprinkler heads that were positioned 3 ft above the top modules in the stacks were melted and released water almost immediately to bring the fire under control. The fire in the second test travelled upward slowly, taking longer to melt the sprinkler heads. As it burned upward, the fire spread through the pallets in the stacks to reach loads in the adjacent row. The comparative burn area analysis indicated much greater damage in the block of non-film-wrapped loads on pallets. The pallets had not only added combustible fuel to burn, but provided channels through which the fire could spread.

It was concluded that the sprinkler system that had been designed to control fires in storage blocks of palletized products would be adequate for palletless, film-wrapped, unitized storage conditions. The tests had highlighted the potential fire hazard inherent in the storage of products on wooden pallets.

With the fire hazard issue put to rest, the next step was the development of machinery to automatically set up and prepare the shrink-wrapped loads. One method involved a machine to position the necessary cardboard sheet on top of the load and then to pull a large bag down over the load. The load on a pallet was taken by a fork truck to a conveyor that carried it through a heat chamber to shrink the film. Upon emerging from the heat chamber, it was conveyed into a large rotating drum that turned the entire load upside down for removal of the pallet.

A second, fully automated system, called Omni-Wrap, was eventually developed by American Can Company and General Foods for application on the Jell-o products packaging line at Dover, Del. With this system, the paperboard base sheet was automatically positioned under each load, and the load was passed through horizontal and vertical film wrapping stations and on through an in-line heat tunnel. In later years, the system was converted to apply the film by the more cost-effective stretch process, which eliminated the need for a heat tunnel.

Shipping posed a key problem for clamped loads. To deposit a clamp load into a transport vehicle, space must be left on either side of the load to permit opening and withdrawing the clamp pads. This space must then be maintained in transit so that clamp pads can access the loads for removal at the receiving terminals. A variety of methods of maintaining the void spaces were tried, but it was difficult to keep loads from shifting during transit. Plugging the voids with paper dunnage was tested. This minimized shifting, but added the costs of dunnage and labor to place and remove it.

Clamping today is widely used for the inside storage of many kinds of products, but it has achieved only limited applications for unitized shipping. Some shippers have combined clamp and pallet methods by using clamps to high stack loads onto a single pallet base for shipment, thereby minimizing the number of pallets required.

Large, bulky boxes of lightweight products have been shipped successfully by the clamp method. The front edges of the clamp pads are used to push the light loads into position to allow access for their removal at receiving docks.

It was eventually concluded that clamp shipping would not be a suitable unitized shipping alternative to pallets for the broad range of product types in the food industry. The search continued for a cost-effective palletless alternative.

Clamp attachments consist of two metal pads with rubber linings. The dimensions of the pads are tailored to suit the range of dimensions of the transport modules to be clamped (Figure 4.1).

The pads are opened and closed hydraulically. The amount of line pressure that is required to clamp and lift a module depends upon the size and the weight of the module. To prevent overclamping and damaging a module, a pressure-relief valve is installed in the hydraulic line. Light modules may require less than 1000 psi line pressure to grip and lift, whereas heavier modules will require more than 1000 psi. It is necessary to adjust the line pressure-relief valve to suit the range of module types to be handled.

5

Unitized Shipping on Slipsheets

The slipsheet concept can be simply demonstrated by placing a small carton on a sheet of paper on a table with the paper protruding out from under an edge of the carton. Grasp the protruding tab with your fingers and pull it along the table. The weight of the carton creates friction between it and the sheet, and it moves along with the sheet. Place your free hand flat in front of the carton and pull the sheet with the carton onto it. Move the carton to another part of the table and pull your hand out from under it while keeping it in place with the other hand.

The earliest known development work on the slipsheet method was done at Clark Equipment Company at Battle Creek, Mich. in 1946. Clark, a major manufacturer of powered lift trucks and lift-truck accessories, had been asked by a client to construct a pusher device attachment for a lift truck to push unitized loads of bags off pallets. The idea was to eliminate handling each bag during loading operations.

Clark engineers designed special lift-truck forks with hook devices to lock a pallet in place as a pusher plate was extended by a hydraulically powered pantograph to push the load off. The mechanical device worked well, but the bags in the bottom tier were damaged, as they were pushed over the pallet edge. Special pallets of minimal height and pallets with rounded edges were tried, but the condition persisted.

Subsequent experimentation with the use of a thin metal plate in lieu of a pallet determined that the bags could be pushed off without damage. The problem was getting the bags unitized on the plate to begin with. A Clark design engineer came up with the idea of unitizing the bags on a sheet of high-tensile kraft paper. One end of the sheet would be left protruding out from under the load so it could be mechanically gripped and pulled with the load onto the thin metal plate. A prototype lift-truck attachment was designed that consisted of a thin metal base plate and a vertical pusher plate to which an alligator jaw gripper bar was installed along the bottom edge.

The first demonstration of the prototype was performed in the plant yard of the old Clark plant at Battle Creek in 1946. Two unitloads of cased goods were positioned on two large sheets of kraft paper. The lift-truck operator extended the pusher and used the gripper bar to clamp onto the section of the sheet that protruded out from under one of the unitloads. He then retracted the load onto the thin plate. Next, he raised and positioned it over the other load and lowered it to contact the top of the load. With the gripper bar released and the pusher plate extended, he slowly backed the truck to pull the thin base plate out to allow the entire load to drop onto the top of the lower load. While much remained to be done to perfect the mechanical system and sheets, the demonstration proved the feasibility of handling heavy loads on thin sheets of paper. Clark patented the device and marketed it under the trade name Pul-Pac. It was introduced as a new and innovative materials-handling system at the first National Materials Handling Show held at Cleveland, Oh. in January 1948, but it attracted little interest. This was probably due to the heavy cumbersome Pul-Pac attachments and poor-quality sheets that were available then. The method was considered during initial research for a unitized shipping method in the food industry in the 1950s. Wooden pallets were much simpler to adapt to unitized shipping at the time. The problems of reusable pallets and pallet exchange were not yet known.

Variations on the sheet materials and the design of the lift-truck attachments to handle sheets appeared within a few years. The term *slipsheet* became the most commonly used name for the concept. The descriptive term *push-pull* became the most popular name for the lift-truck attachments. (see Figure 5.1)

The early commercial applications of the slipsheet method did not involve the replacement of pallets. They offered an alternative to nonunitized hand methods where pallets were not being used. The use of pallets for car and truckload shipments of lightweight products, in particular, were cost prohibitive due to transportation penalties. The space the pallets took up under each unitload, (over 6 cu ft), would reduce net payloads and increase the cost per net ton shipped.

The long-term unitized storage of seasonally produced products, such as processed fruits and vegetables, were also not economically suited to pallets. These kinds of products are produced in relatively short harvest seasons and warehoused for distribution throughout the year. An enormous investment in pallets would be required for the seasonal build-up of inventory. Palletless unitized storage on slipsheets offered a more cost-effective alternative.

The use of pallets for the shipment of raw materials and other goods that did not enjoy the free return privilege of exchange system pallets opened further opportunities for slipsheet applications. The transportation costs to return pallets for reuse often was substantially more than the costs of slipsheets.

With the deterioration of the pallet exchange system in the 1970s, manufacturers in the packaged foods industry focused attention on the slipsheet method as an alternative to wooden pallets, particularly for the long hauls from plants to regional distri-

Figure 5.1 A lift truck with a push-pull attachment. The carrying base of the push-pull attachment consists of two flat metal plates, each 15–18 in. (381–457 mm) wide and 48 in. (1219 mm) long. There is a 4.0 in. (102 mm) gap between the plates. The attachment is used to carry slipsheeted loads in the 48×40 in. (1219×1016 mm) size range. Optional carrying bases include multi-tines (a series of six or more narrow forks) and single wide plates. The slipsheet gripper bar is located along the base of the pusher plate. The pusher plate with the gripper is extended and contracted hydraulically on a pantograph from the mast of the lift truck. A locking device is used to stop the load at the edge of the 48 in. long plates when carrying loads less than 48 in. deep. This prevents the tips of the plates from cutting into the load ahead when depositing or retrieving modules. Photo courtesy of Cascade Equipment Corp.

bution centers. In the beginning, many management people had concerns over the feasibility of adapting the slipsheet method to widespread use throughout the industry. The key concerns included:

Cost justification for the purchase of lift-truck push-pull attachments to handle slipsheets. In many instances, existing lift trucks would have to be ungraded to carry both loads and attachments.

Potential productivity losses that could result in additional lift trucks and operators. Demonstrations of the slipsheet method indicated that, in comparison with the relatively simple pallet method, slipsheets would be a slower and more difficult method.

Receiver acceptance of the slipsheet method. It was questionable if the intermediate consolidation terminals would make the investment in push-pull equipment that would be necessary to receive the manufacturers products on slipsheets.

Durability of slipsheets for transport of unitloads through a complex system of transfer terminals in which slipsheet tabs would have to be grasped and pulled a dozen or more times.

Adaptation of the slipsheet method to mechanical handling equipment such as unitizing machines. Automatic pallet loader machines were in common use on most production lines. Slipsheets would have to be inserted under the loads.

Unitload preparation and the packaging techniques would have to be changed. The "bite" of pallet boards while sometimes causing damage to shipping cases, held the loads together for materials-handling operations. Loads had to be stabilized on the smooth-surface slipsheets and held in place so they would not encroach on tab scorelines during shipment.

Because of the unknowns at the time, most companies ran lengthy tests, sometimes for a year or more, in order to identify problem areas that needed to be corrected in order to have viable cost-effective systems. Many improvements to slipsheets and related handling equipment and unitizing methods were made during that time.

Push-pull attachments were substantially improved. New lighter weight attachments that allowed improved operator visibility of the loads were developed. Attachments with double blades that could be spread or contracted made handling of either pallet-loads or slipsheets possible without changing from fork to push-pull attachments.

Paperboard slipsheets, too, were improved. More durable tabs were developed, and tab ends were rounded or cut at angles to facilitate maneuvering without damaging other products. Moisture-resistant sheets became available so that damp shipping docks or humid storage areas would not damage nor weaken them.

Decisions had to be made on the size and type of slipsheets to use. Cheaper corrugated board with laminated outer sheets of kraft paper were adapted by a number of shippers. Quality assurance people at one large food company ruled out the corrugated sheets on the basis that corrugations provided channels for insect infestation. Solid fiber sheets were the only acceptable alternative to them. The thickness of sheets depended on the weight of the load and the strength needed.

Dimensions were generally based on the standard 48×40 in. pallet, since for many years unitized modules had been sized to fit the 48×40 in. pallet surface as closely as possible. The dimensions of slipsheets were not as critical as those for pallets, since there were no sharp edges for products to overhang. There was concern, however, that unitized modules on slipsheets fit into standard rack bays and warehouse slots designed for the 48×40 in. pallet. Modularity to the inside clearances of transport vehicles further dictated care in establishing slipsheet dimensions. Shippers and receivers differed on their preference for slipsheet tab width. Some preferred a wide tab of 4 in. (10 cm) to facilitate mechanical gripping. Others felt a tab that wide would crimp and hamper the gripping operation. A very short tab, 2 in. (50 mm) or less, on the other hand, would be difficult to get entirely into the gripper jaws. Today a commonly accepted tab width is 3 in. (76 mm).

To stabilize products on the sheets, some manufacturers installed glue dispensers or sprayers on their production lines. The glue would be dispensed as stripes, as spots, or sprayed onto the top flaps of shipping cases, and also onto the top surface of the slipsheet. This helped to stabilize the loads and hold them together as a unit.

With the availability of stretch film and stretch-film machine applicators in the 1970s, many shippers elected to film-wrap loads on slipsheets. A number of different kinds of stretch-film wrapping machines became available in the 1970s. The kind of wrapping machine used depends upon the number of unitloads produced per hour. Spiral wind machines are commonly used today. One type involves rotating the unitload on a turntable as the film is unrolled and wrapped around the load. Another method is to hold the unitload in place as a machine robot arm wraps the film band tightly around the load. Very high volume production lines may use a plow-through machine wrap method in which the unitloads move along a conveyor through a web of film that is pulled tightly around each load and tack sealed in place.

Infrequent or very low volume production may involve the use of simple hand applicators in which a handheld spool of film is unwound and applied by walking around the load and pulling the film tightly around it.

The film wrapping of unitized transport modules adds to their costs, but it is usually justified by the reduction of damaged goods and improved handling efficiency. It essentially converts a group of loose containers into a single, large container for more efficient mechanical handling throughout the system.

While its use may be critical to palletless handling efficiency, it is also effective in containing loads on wooden pallets, and it is considered by many shippers today to be a necessary system cost for any kind of unitized shipping.

The slipsheet method should be regarded as a system and not just an alternative unitload transport base. Preparation of the slipsheet module is a packaging as well as a materials-handling function. Conversion to the slipsheet method from either manual or palletized methods requires a management commitment of time and resources, and an investment in equipment.

Depending upon the specific application, the investment in equipment could be substantial. Machinery could include not only the lift-truck attachments to handle the sheets, but in some cases the replacement of the existing lift trucks with new ones to carry both the attachments and the loads.

Machines to position slipsheets onto pallets, and machines to transfer loads to and from pallets at shipping docks, may be required for high-volume operations. If film-wrap equipment is not already available, it too may have to be installed as part of the slipsheet system cost.

In general, management people look upon such expenditures as an investment that must be cost justified by the savings it generates in comparison with existing methods of handling and shipping. If nonunitized methods are currently used, the elimination of the labor costs to manually load and unload each piece may justify the investment in the new system. The volume shipped is a critical element in establishing the amount of savings and cost justification for equipment.

If the cost comparison is made to an existing palletized shipping system, it involves identifying the savings generated by the elimination of pallets and the miscellaneous costs incurred by the use of pallets. The determination of pallet costs, other than the initial purchase price, is not a simple matter. In a pallet exchange system, for example, the useful life of an average pallet is usually a guess on the part of users. Identifying the total costs on an annual basis is equally difficult, since the lifetime costs of a pallet in an industrywide exchange system may be distributed over many cost centers.

To further complicate the comparative analysis of pallet versus slipsheet system costs, the indirect costs caused by the use of pallets may be well hidden in the costs of other activities in the system. As an example, space taken up by pallets under the loads can limit the amount of product shipped and add to transportation costs. Just how much would depend upon the particular product and would be reflected in incremental transportation costs per net ton shipped. Another hidden cost is that of the damage to products that is traceable to pallets. An analysis sheet (Figure 5.2) lists the cost areas usually considered in a pallet versus slipsheet comparative system cost analysis.

	PALLETS	SLIPSHEETS
Total purchased annually		
Initial price each		
Annual total cost to purchase		
Procurement costs — shipping, receiving, handling, sorting		
Return costs/year for reuse		
Repair costs/year		
Annual depreciation — equipment		
Annual maintenance — equipment		
Annual transportation costs		
Annual labor costs		
TOTAL ANNUAL COST OF SYSTEM		
# OF UNITIZED LOADS SHIPPED/YEAR		
COST PER UNITLOAD SHIPPED		

Figure 5.2

Before attempting to put such an analysis together, both present and proposed systems should be flow charted in detail so that all cost areas applicable to each system are identified. In this process of research, it may be determined that a composite system is best suited to a particular application, such as the use of slipsheets for the long-distance transport of products, with pallets used for local distribution or for storage.

During the 1970s many large manfacturers who had been shipping their products totally on wooden pallets converted substantial volumes to the slipsheet method.

A cost-justification summary of a conversion plan for a large manufacturer is shown in Figure 5.3. This project called for the elimination of 500,000 pallets from the company's North American shipping operations.

Pallet replacements and repairs were costing $2,340,000 an-

PROJECT—ELIMINATE 500,000 PALLETS FROM DISTRIBUTION SYSTEM

Current annual pallet purchases --------------------	$2,340,000
Added costs incurred by use of pallets -------------	580,000
TOTAL ------------------------	$2,920,000
Investment required for slipsheet system ----------	$1,200,000
Annual costs of slipsheet system:	
Purchase paperboard slipsheets ------------	$1,500,000
Incremental depreciation and	
maintenance costs ----------------------------	400,000
Contingency ----------------------------------	200,000
TOTAL ------------------------	$2,100,000
Projected annual cost savings -----------------------	$ 820,000
Potential transportation savings --------------------	$1,500,000

Figure 5.3

nually. Related costs, including the administration of the pallet control program, the procurement and shipping costs of new pallets, the cost of lost pallets, and the costs of relocating empty pallets to balance shipping and receiving needs added $580,000 dollars to the annual cost, for a total of $2,920,000.

The conversion to a slipsheet system required an investment of $1,200,000. Annual purchases of fiberboard slip sheets amounted to $1.5 million dollars. Included in the system costs was the depreciation and maintenance of push-pull attachments for lift trucks at 35 shipping and receiving locations. A contingency was budgeted to cover unknowns.

The projected cost savings that the slipsheet system was expected to produce were adequate to justify the necessary investment in time and equipment to implement the conversion program. The financial plans usually did not include savings that were later realized through improved payloads in transportation equipment that slipsheeting made possible. This potential area of savings was of course a consideration in the financial assessment and in the determination of the risk involved.

The impact of conversion from pallets to slipsheets in the food industry had an impact on shipping modes. Intermodal shipping, that is, shipments in highway trailers that ride piggyback on rail flat cars, provides faster long-haul service, since the time to switch rail cars to receiving locations is eliminated. Neither shippers nor receivers wanted to bear the labor costs to hand load and unload goods in the trailers and were reluctant to ship unitized on pallets, since the pallets would have to be collected and returned to shippers for their reuse at substantial incremental shipping costs. Slipsheets made the unitized loading and unloading of piggyback trailers cost effective, since the sheets did not have to be returned. The faster intermodal service in many cases reduced the amount of field inventory required to service customers.

The impact of eliminating 500,000 pallets from a company's logistical system is illustrated in Figure 5.4. It visualizes stacking that many empty pallets 50 high and placing the stacks in one long line. The result is a line of wooden pallets, 25 ft (7.62 m) high and 7.5 miles (12.0 km) long. Total tare weight is approximately 18,000 tons of wood. It gives an indication of the potential for increasing net payloads with the slipsheet method.

With successful conversion to the slipsheet method for substantial volumes, large shippers turned their attention to reducing the number of pallets still coming inbound to manufacturing plants with unitloads of raw and packaging materials. The pallets were previously recovered and put into the finished goods pallet system. With that need diminished, the number of pallets from inbound materials were becoming a costly nuisance to dispose of at many plant locations. Consequently, many manufacturers who received truckloads of palletized raw materials now preferred to receive slipsheeted loads. The cost savings that represented the difference between pallets and slipsheets were usually shared by the supplier and the receiver.

Where volume did not justify the immediate purchase and installation of push-pull attachments on the raw material shipping and receiving docks, special thick slipsheets were sometimes used

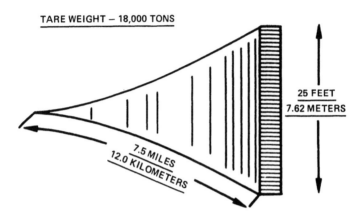

Figure 5.4 The impact of eliminating 500,000 pallets standard food industry pallets from a large food manufacturer's finished products distribution system in the 1970s is illustrated in the above drawing. A standard industry pallet is dimensioned approximately 48×40×6.0 in. (1219×1016×152 mm). It takes up approximately 6.7 cu ft (0.19 cu m) under the unitload. Place 500,000 empty pallets in stacks 50 high each and place the stacks adjacent in one long line. The line of wood is 25 ft (7.62 m) tall, and extends 7.5 miles (12 km) in length. Tare weight of that many pallets would be approximately 18,000 tons. Palletless unitized shipping on slipsheets provides an opportunity for the shipper to get more product into the transport vehicles, thereby reducing the shipping cost per net ton. That is the major reason for the use of slipsheets as an alternative unitized shipping method.

under the loads. Tapered fork tines could chisel in under the thick sheets and transfer the loads to pallets on the shipping docks. As slipsheet volume grew on the raw material docks, push-pull attachments could be economically justified by the savings in the cost difference of the special thick sheets and conventional slipsheets.

It was during the research that was necessary to convert raw material deliveries from pallets to slipsheets that the possibility for large bulk boxes and bags became apparent. It was found that truckload lots of many materials could be packaged in large, single bulk containers that held the equivalent contents of an entire slipsheet load of the smaller containers. The development of these large palletless containers for raw materials is discussed in Chapters 11, 12, and 17.

The increased use of slipsheets throughout the United States resulted in the need for the standardization of the definition of slipsheets as well as their materials and testing procedures. The Physical Distribution Standards Management Board of the American National Standards Institute in 1978 assigned the task of the development of a slipsheet standard to the Pallets Standards Committee of the American Society of Mechanical Engineers. Up to that time the only standard for transport modules was based on the use of pallets. It was identified as ANSI standard MH1.

A committee was subsequently formed of representatives from a broad cross section of manufacturers of slipsheets and related equipment, as well as shippers, the transportation industry, and government agencies. The committee established two working groups, one to be concerned with the definition and sizes of slipsheets, and the other with the materials and testing procedures. The title of the existing MH1 pallet standard was changed to MH1.5M and titled: "Standardization of Pallets, Slip Sheets and Other Bases for Unit Loads."

The standard that was approved on November 5, 1980 identifies slipsheets as a flat sheet of material with tabs for the handling and transport of unitized products. The materials of which slipsheets are made are identified as corrugated fiberboard, solid fiber paperboard, and any combination of polymerized materials such as polyethylene or polypropylene plastics. The standard includes guidelines for the kind of slipsheet to use for different kinds of products and shipping conditions. the ANSI MHI.5M standard was published by the American Society of Mechanical Engineers. Copies can be purchased through the American National Standards Institute, 1430 Broadway, New York, N.Y. 10018.

6

Bulk Container Transport Modules

A bulk-container transport module is a large box or bag that contains the equivalent amount of goods or materials in one or more unitloads of smaller containers. Some examples would be a wooden pallet bin to transport field crops from a farm to a packing plant, a large fabric bag to carry dry bulk materials, or a large paperboard box to transport and ship bulk goods or materials of many kinds (Figure 6.1).

Bulk containers, like unitloads, eliminate the slow and tedious handling of many small containers and speed the packing, transport, and discharge of contents with the use of mechanical handling methods and equipment. The large bulk containers have been given a number of trade names, such as IBCs (intermediate bulk containers), pallet boxes, pallet bins, bulk tote boxes, heavy-duty containers, and industrial containers (Figure 6.2).

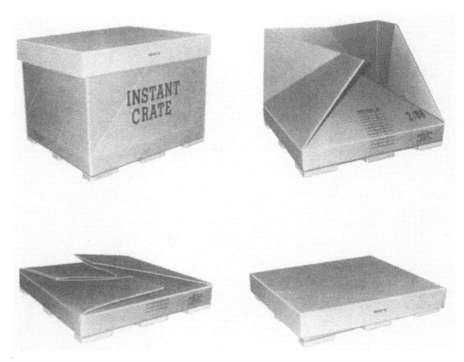

Figure 6.1 Corrugated crate with glued on plastic squares for handling by fork-lift tines. Photo courtesy of Containair Systems Corporation, New York, N.Y.

The development and introduction of large wooden pallet bins for the transport of produce from fields and orchards to packing plants was one of the earliest applications of bulk containers. It began with research carried out by the University of California at Davis, Cal. in the mid 1950s. Up to that time the traditional methods of harvesting field crops involved the picking of fruit in buckets and transferring it into small lug boxes. The lug boxes were manually loaded onto flatbed wagons for transport to the packing plants. There they were unloaded by hand and the contents dumped either into receiving hoppers or onto conveyor belts for processing or packing. Certain changes were taking place in the 1950s that gave impetus to experimentation with bulk bins and

Figure 6.2 Corrugated bulk box with integrated slipsheet. Photos courtesy of Containair Systems Corporation, New York, N.Y.

mechanized methods of handling produce. New roads and highways were being completed that would make it possible to carry freshly harvested produce longer distances on bigger vehicles to larger central packing plants. Rapid turnaround of the transport vehicles was essential to handle the growing volumes. To speed the loading and unloading operations, lift trucks were acquired for the plants and fork-lift attachments were installed on field tractors to handle palletloads of lug boxes.

The availability of lift-truck equipment to handle pallets made it possible to experiment with large bulk bins that could substantially improve productivity and reduce the labor costs associated with the manual handling of lug boxes. However, a great deal of research and development work was required to design suitable pallet bins that could be integrated efficiently into the handling and shipping process.

In 1956, a prototype pallet bin for fruit was designed at the University of California at Davis, Cal. It was 47 in. (1194 mm) square and 24 in. (610 mm) deep inside. A wooden pallet base was integrated with it, and the pallet's top boards served as the flooring of the bin. The side walls of the bin were constructed of 0.625 in. (15.9 mm) plywood panels, which were attached to 4×4 in. (102 mm × 102 mm) upright corner posts. The bin was designed for a capacity load of 1000 lbs (454 kg).

An early concern was that fruit that travelled many layers deep in bulk bins would be more vulnerable to over-the-road damage than fruit in shallow lug boxes. Comparative field tests, however, determined that less damage occurred to fruit in the bins than in an equivalent number of lug boxes during the over-the-road hauls. Tests made with cling peaches at the University of California established the reason for this.

It was found that most of the damage in transit occurred in the top layers of fruit in the containers. The fruit at the top was not stabilized by the weight of the load and could be tossed about freely during the trip. As vibrational forces approached a g force of 1.0, the pieces on top were periodically rendered weightless, which allowed them to rotate and strike one piece against another, resulting in bruise damage. The percentage of fruit in top layers in relation to the total contents was much less in bins than in lug boxes. Consequently, the results of comparative testing showed a significantly lower incident of bruise damage to the fruit in the pallet bins. The studies further determined that the stabilized fruit deep inside the bins was much more resistant to compression damage than had been anticipated.

Considerable research and development work on pallet bins was carried out by the United States Department of Agriculture at the Forest Products Laboratory at Madison, Wis. in the 1960s. Design considerations established for the wooden pallet bins at the time included:

Standardization of bin sizes was essential to accommodate automatic dumpers, stackers, destackers, and other materials-handling equipment at processor plants.

The length and width dimensions were designed to be modular to the flat-bed trucks and trailers of the time. The maximum width allowable for highway vehicles then was 96 in. (2438 mm). Standardization on the 47 in. (1194 mm) OD bin dimensions permitted two bins to be placed adjacent on the bed of the vehicle, which would still allow space for tie-down ropes without exceeding the maximum width limit.

With length and width standardized at 47 in. (1194 mm), the height could be varied according to how vulnerable the particular fruit would be to crushing. For example, 30 in. (762 mm) was found to be maximum for apricots and peaches, 42 in. (1067 mm) for apples, and 35 in. (889 mm) for potatoes.

The cubic capacities were generally governed to accommodate a total of 1000 lb (454 kg) of most fruits. The bins that were stacked two high and two adjacent on flat-bed vehicles distributed the load evenly over the decking for efficient highway transport.

Wood was the ideal packaging material for the construction of the bins. It was a strong but inexpensive material. If the bins were properly designed and maintained, they could be expected to last many harvesting seasons with little upkeep or repairs.

Although the tare weight of the bins ranged from 80 lbs to over 150 lb (36 kg to over 68 kg), the impact on transportation costs was not that great, since the hauls from field to plant were relatively short. Wood made it possible to build stacking strength into bins so that they could be floor stacked up to 10 high to make good use of costly controlled atmosphere and refrigerated storage space. Empty bins could be stacked outdoors in huge storage blocks during the off season (Figure 6.3).

By 1960, hundreds of thousands of wooden pallet bins were in use throughout California, Washington, and Oregon. Farmers and processors all over the country were changing to the pallet-bin bulk container systems. Bulk bins were so successful that some

Figure 6.3 Stacks of empty pallet bins await the harvest season for pears and apples at a Hood River, Or. packing plant. During harvest season each bin is filled with approximately 1000 lbs (454 kg) of fruit. The fruit remains in the bins in controlled atmosphere storage rooms and is processed and packed for shipment throughout the year.

shippers envisioned using them for the transport of produce to distant wholesale market centers and possibly all the way to large supermarkets.

There were, however, two major drawbacks to the use of the bins for longer hauls. First, the tare weights of the bins would contribute to higher net shipping costs. Second, the bins would have to be reused to be economical. Since they could not be collapsed or nested for the return haul, the shipping of empty bins long distances was cost prohibitive.

As corrugated paperboard technology progressed, large bulk paperboard containers could be designed to transport a thousand pounds (454 kg) of bulk product. The storage and transport of loose frozen (IQF) vegetables was among the first applications for the boxes. IQF is a process in which products such as peas, or cut pieces of carrots or other vegetables and fruits, are frozen on a moving belt. Each piece is thereby frozen separately, and the result is a flowable mass of frozen bits. Many field crops today are loose frozen during the harvest season and are shipped in bulk to packing plants in market areas for retail packing as needed the rest of the year. Before bulk paperboard boxes became available, the loose frozen bits were packed in small cases, which were hand stacked into reefer cars and trucks for shipment to the regional packing plants. There, the cases were opened by hand and the contents dumped into the receiving hoppers on the packing lines.

The large boxes were either cap and tube style or RSC (regular slotted container) type with full flaps top and bottom. They were set up on pallets and fitted with polyethylene liner bags. The outside length and width dimensions matched the 48×40 in. (1219 mm×1016 mm) pallet surface. The overall height of the boxes on standard pallets was approximately 36 in. (914 mm) to provide cubic capacity for an average of 1000 lb (454 kg) for most vegetables. It was found necessary to strap the boxes onto the pallet bases to keep them from shifting over the edges of the pallets during shipment. To discharge the contents, the boxes were placed onto mechanical devices that tilted them to allow the contents to flow out of the tops by gravity. After use, the empty boxes were removed from the pallets, collapsed, and strapped into bundles for return to the packing plants for reuse the next harvest season. Typically, the boxes lasted two or three harvest seasons before they had to be discarded.

Pallets simplified materials handling but were not the ideal carrying base for bulk containers of IQF product. They added approximately 85 lb (38.6 kg) of wood to maintain under each load at freezing temperature during the trip. They took up 6 cu ft (0.17 cu m) of space under each box, which reduced the net payloads and added to transportation costs.

As slipsheets came into wider use in the 1970s, large shippers began testing the use of paperboard slipsheets under the bulk boxes in the place of pallets for their shipments. The sheets were usually placed on pallets at the packing plants, and the empty bulk boxes were set up on top. Filled boxes were transported through storage areas to shipping docks on the pallets. The slipsheets were employed for the outbound loading operations. At the receiving locations the boxes were removed on the slipsheets and transferred back onto pallets for storage and for later transport to packing lines. The successful application of paperboard bulk containers for IQF products stimulated interest in their use for many other kinds of goods and materials.

The economic justification for containerized shipments of raw materials is related to the kind of material and the volume shipped. Materials such as sugar, wheat, soybeans, and a few other agricultural and chemical products are commonly shipped in enormous tonnages, which justify the use of total bulk shipping methods with dedicated transportation and materials-handling systems. Materials shipped in bulk hopper rail cars and trucks, or liquids shipped in bulk tankers, eliminate the need for containerized packaging.

Most materials today, however, are packed and shipped in containers of some kind. These may be steel or fiber drums, corrugated cases, small wooden crates or boxes, burlap sacks, or plastic and paper bags. In the early 1980s, with shipping volumes increasing at a steady rate, many of these smaller containers were being shipped in full truckload, carload, or ocean container quantities. At the request of the receiver locations, many suppliers shipped the containers unitized on standard pallets. When unloaded, these pallets could be fed into the finished goods distribution system. With the large-scale conversion to slipsheets for the finished-goods long hauls, the need for pallets was substantially reduced, and consequently pallets from the inbound loads of raw materials had to be disposed of in other ways or sold as scrap lumber.

One solution was to get suppliers to ship raw materials unitized on slipsheets instead of pallets. Integrating the use of slipsheets with manufacturing operations was difficult at the manufacturing plant locations, since few of the lift trucks that operate around the plants could be adapted to carry push-pull attachments. Low-cost pallet jacks and trucks were used at many locations to transport pallet loads from docks and holding areas to the production lines. Therefore, slipsheet modules of raw materials arriving at the plant shipping docks had to be transferred onto pallets for plant handling. Thick, heavy-duty slipsheets were used under some loads to enable lift-truck fork tines to chisel in under the loads to lift and transfer them onto pallets. That saved the cost of installing push-pull attachments on fork trucks on those docks where the volume of slipsheeted loads received did not justify the investment in such equipment.

The opportunities for bulk container packaging of many materials that were traditionally received in smaller containers became apparent during studies of ways to ship raw materials without pallets.

It was found that some materials shipped in burlap sacks or cloth and paper bags could be converted to containerized bulk handling in large, plastic fabric bags. These bulk bag containers are handled palletless by the use of top straps. They are loaded through the top and their contents discharged through the bottom of the bag.

Bulk bags made of woven polypropylene fabric became available in the 1970s. They added very little tare weight to the loads, and they were collapsible into very small packages for return and reuse. However, they had certain functional drawbacks that limited their application to certain kinds of materials and storage conditions. They were difficult to store more than two high in warehouse storage areas. The materials had to bear the total load pressure, since the fabric container walls contributed no top-to-bottom compression resistance. Fragile materials, therefore, could be crushed and damaged during handling and storage. Finally, the materials had to have sufficient lubricity to accommodate discharge by gravity flow. Chapter 17 contains a case history on the application of bulk fabric containers.

While large paperboard bulk containers on slipsheets were effective for shipment of IQF bulk products, many design improvements were needed before they could be adapted to wider scale use for other products. The slipsheets had to be glued onto the bottom flaps of the container in order to stay in place during shipment. A method of integrating the slipsheet with the bottom of the container was needed. Also, the RSC style for this size container was unwieldy to handle and usually required two people to set up the box for filling. Stackability of paperboard containers in humid storage areas was still another concern. Chapters 11 and 12 contain case histories on the development and design of large paperboard containers to accommodate the slipsheet method of handling.

7

Materials Handling

The size and weight of transport modules require that they be handled mechanically throughout the system. The designer of transport modules should be familiar with the kinds of materials-handling equipment used for the handling and the transport of the modules at the shipping and receiving locations throughout the system.

The materials-handling functional activities that apply to transport modules will include the following:

BULK CONTAINERS OF MATERIALS

Transport of empty containers and liner bags from receiving docks or storage locations to the filling stations

Set-up of containers, installation of liner bags, filling, densification of materials, and weighing

Preparation of the container for shipment — closing and sealing, film overwrapping, strapping

Transport of the containers from filling stations to storage or holding areas, and from there to the shipping docks

Loading the containers into the transportation vehicles and securing them with dunnage if required

Unloading the containers at receiving locations and transporting them to storage or holding areas

Transporting the containers from storage or holding areas to the discharge stations

Preparing the containers for the discharge of loads

Discharge operations

Preparation of waste packaging for trash disposal or recovery of reusable containers and pallets for reuse

UNITIZED LOADS OF SMALLER CONTAINERS

The assembly of the small containers into unitload patterns

Preparation of the transport module for shipment such as overwrapping with stretch film or strapping

Transport of the modules to storage or holding areas

Transport of the modules to shipping docks

Loading the transportation vehicles at the shipping docks, including use of dunnage if required

Unloading the modules at the receiving locations and transporting them to storage or holding areas

Transporting the modules to breakbulk locations

Depalletizing or breaking down the modules piece by piece for delivery of contents to final destination

Trash disposal of the disposable materials such as wrapping, strapping, and dunnage materials

Recovery and preparation of reusable materials such as pallets

The amount of packaging time and resources that is prudent to spend on the development of packaging to improve the productivity of materials-handling activities will depend on the shipping volumes. Where substantial volume exists, it is important that the developer of the transport modules spend as much time as necessary to thoroughly research and build knowledge on the kinds and costs of materials-handling equipment that are available. It may be necessary for certain projects to specify the specific type of materials-handling equipment that will be needed in order to handle the type of modules that will be shipped or received.

The mechanical handling equipment that applies to the materials handling functions can be summarized under three general types:

Mobile handling equipment: the equipment used to physically move the modules from one location to another.

Module preparation equipment: the equipment used to assemble the unitloads or fill the bulk containers.

Delivery end equipment: the equipment used to break down unitloads or to remove the contents of large bulk containers.

Just about every major project will have different needs for materials handling and therefore different types of equipment. Flow charting the system through which the modules will pass will help identify the specific materials-handling requirements. The most common kind of equipment will be discussed in this chapter.

MOBILE EQUIPMENT

The most common equipment found on shipping docks, in production plants, and in warehouses is the fork-lift truck. In flow charting a system it is important to make notes on the types of lift trucks and their load capacities. Also, note the kinds of handling attachments available, such as clamps, push-pull, top-lift, or strictly fork tines. If the system calls for the handling of palletless loads on slipsheets, it is important to know if all lift trucks that will handle the loads along the route are either equipped with push-pull attachments or are capable of being equipped with them.

Most lift trucks that have been put into operation since the introduction of palletized shipping are equipped to handle the standard 48×40 in. pallet with loads up to a ton each. The ratings on these trucks usually are sufficient to carry the palletloads two high, or up to two tons. Higher or lower rated truck capacities may be found at specific locations where only very heavy or very light loads are normally handled.

In most locations, the types of lift trucks can be narrowed down to two general classes, the straddle type and the counterbalance type. It is imporant to know something about the physical characteristics of these different types of trucks in order to check out their suitability to handle large transport modules.

Functionally, the straddle trucks may not be as easy to operate, nor as universal in application, as the counterbalance trucks, but their sharp turning radius permits them to be operated in relatively narrow aisles and in spaces where access is limited.

The straddle-type trucks are battery powered. The operator usually stands on a rear platform in back of the battery compartment. Two outrigger arms with front wheels extend from the mast and serve to stabilize and bear the weight of the load lifted. The module must be dimensioned to fit easily within the space inside the outrigger arms.

The most common type of lift truck is the counterbalance truck. They can be either gas powered or battery powered. While some stand-up models exist, most are operated with the rider seated. The traction wheels are usually on the front axle, although there are models that use a combination traction plus a steering wheel in the rear for the purpose of narrowing the turning radius. The load is carried on the fork tines or other types of attachments that extend out from the mast. The front axle is the fulcrum, and the weight of the load is counterbalanced by the weight of the truck and the battery if it is battery powered. The lifting height and added functions, such as side shifting and mast tilt, will also have an effect upon the load capacity. The load ratings are shown on a metal plate that can be found on the front of most trucks.

Information about the lift trucks that will be pertinent to the size and design of transport modules includes:

What is the maximum load that can be safely lifted and transported? Plates on the trucks provide the horizontal and lift load ratings at fixed load centers, commonly 24 in. (61 cm). If the module being proposed exceeds the 24 in. load center rating, or if the plate cannot be found on the truck, check with the engineering department or the manufacturer of the truck to ensure that the module being proposed can be safely handled.

If there exist options on the size and weight of a module, the truck load rating may influence the module size selected.

If the truck is an outrigger type, what is the clearance required inside the outrigger arms?

What is the maximum extended height and the collapsed height of the lift mast? Masts are designed to telescope during the lifting process in order to stack high enough to utilize overhead storage space and to collapse low enough to permit passage through doorways of rail cars and trucks at the shipping dock.

How much free lift is there in the mast? This is the maximum height to which the tines or other attachments can be lifted before the mast begins to telescope upward. This can be important if there is a need to stack one module on top of another in a vehicle or storage area where limited overhead clearance exists. Some masts are designed to telescope as the load is raised and have little or no free lift. The height of the module plus another 4 in. should be specified for free lift in most operations. For example, assume the container is 45 in. tall. In order to raise it for stacking on another 45 in. high container without telescoping the mast, you would need at least 49 in. of free lift.

Does the truck have a side shifter? This allows the operator to shift the load laterally 4 in. (10 cm) to either side. This is important to ensure that the lift-truck operator can maneuver modules into tight-fitting slots and to compress modules together during loading operations.

What is the measurement of the wheel base? This is the distance between front and rear axles. If the truck has not been used to load transport modules in the past, it will be necessary to specify the use of a dockplate to bridge the gap from dock to decking of the transport vehicle. The axles of the lift truck should both be horizontal on the dockplate to permit entry of the truck with a high load through the doorway.

What kinds of attachments are used with the truck? Standard fork tines are usually 42 in. in length, 4 in. in width, and tapered from 0.5 in. to 2 in. or more at the base. If forks are too long for the module width, they can penetrate modules in front when depositing a load. If not long enough, the module can tilt over the ends of the tines, which can crease or damage the bottom of the load. If the module extends far over the ends of the tines, there is a danger of it falling off.

If push-pull or clamp attachments are used with the truck or may be used when the module system is installed, it is important to know the revised safe load rating. The plates on lift trucks generally show the rating with standard fork tines. Push-pull and clamp attachments, depending upon the type, will add several hundred pounds to the load and will reduce the net load capacity. Lift-truck dealers should furnish the revised load rating when installing the attachments.

Does the lift truck have quick-disconnect couplings for a fast change of attachments? This will permit changing over from one kind of attachment to another when needed. For example, the truck may normally use fork tines. If an occasional load arrives with slipsheets, the fork tines can be removed and replaced with a push-pull attachment in 2 or 3 minutes if it is equipped with fast-change fittings.

The most common attachments used with lift trucks include:

two fork tines

multi tines (see Chapter 3)

clamps (see Chapter 4)

fork tines with pusher plate

push-pull attachments that fit over fork tines

push-pull attachments (see Chapter 5)

convertible fork/clamp attachments in which short clamp arms are hydraulically rotated to become wide fork tines

top lift attachments (see Chapter 13)

rotator attachments — to invert containers for the discharge of contents or to accommodate the installation of slipsheets under modules

attachments to support strap-slung containers

It may be necessary for some projects to upgrade the truck equipment to cost effectively handle new transport modules. Assume, for example, transport modules on slipsheets that have a gross weight of 1500 lb (680 kg) will be carried in stacks two high for a total 3000 lb (1360 kg) load on the lift trucks in the system. If it is found that a lift truck equipped with a push-pull attachment will have an inadequate carrying capacity for a 3000 lb load, then two options exist. At a sacrifice of time and productivity, the truck may handle just one module at a time. In order to carry two modules at a time, it will be necessary to replace the truck with one that is properly rated for the job. The decision on investing in a higher rated truck will usually depend upon the frequency and the volume received at the location.

LOW-LIFT HAND TRUCKS (Figure 7.1)

These types of trucks are commonly used in conjunction with pallets or skids for horizontal movement of the modules. The arm extensions of the low-lift truck enter the pallet under the top boards of the pallet or skid, and small wheels at the front of the arms are in contact with the floor. The 48×40 in. standard pallet was designed with gaps between its bottom boards to permit the small wheels in the front of the arms to access the floor. When positioned under the pallet, the arms are raised a few inches over the small wheels by means of a hydraulic jack system. This lifts the load clear of the floor so that it can be transported horizontally.

Relatively low-cost, manual low-lift trucks are used for lighter loads under a half ton. Battery-powered, low-lift trucks are in common use today for low-lift handling of loads up to two tons.

Low-lift trucks equipped with push-pull attachments are in limited use. They are used to load and unload slipsheet loads from trailers and to move slipsheet modules horizontally around plants and warehouses.

The operator usually controls the low-lift truck while walking, and the term *walkie* is sometimes used to identify low-lift trucks. Some, however, are equipped with a small platform and holding bar so that an operator can climb aboard and direct the truck as it travels long distances.

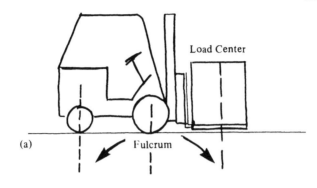

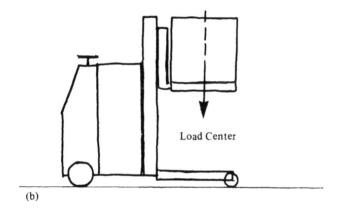

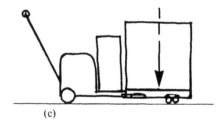

Figure 7.1 Lift truck types: (a) Counterbalanced truck, (b) Outrigger type truck, (c) Pallet low lift truck. Drawings by the author.

PEDESTRIAN TRUCKS

Low-lift trucks should not be confused with pedestrian lift trucks. Such trucks are also operated by walking or by standing on its platform, but they are equipped with lift forks or other attachments so that modules can be elevated and stacked as well as moved horizontally. Such vehicles are used in production areas to transfer stacked modules to and from discharge stations or process areas. They can be equipped with push-pull or clamp attachments. The drain on the batteries of these kind of lift trucks usually discourages their use with heavy attachments, since the batteries will require frequent recharging.

MISCELLANEOUS MOBILE EQUIPMENT

Other types of materials-handling vehicles include low-level platform trucks and tractor-trailer trains for horizontal movement. Mid- to high-rise warehouses operate a variety of stacker systems that range from high-lift rider cranes to fully automated, computer-controlled stacker crane systems. Dimensional tolerances of the modules handled in such systems are especially critical. Bulging or leaning modules are unacceptable. (See Chapter 9 on warehousing for further information on this kind of materials-handling equipment.)

In preparing a system flow chart, the observer should make note of any unique handling equipment along the way that will be used to move the modules. Either the transport modules must be designed to be compatible with the existing materials-handling equipment or the equipment must be upgraded or replaced to be compatible with the packaging system.

UNITIZING EQUIPMENT

The standardization of pallet size in the 1950s was followed by widespread installation of automatic pallet loader machines. A stack of pallets is inserted into a chamber of the machine, and either the top or bottom pallet, depending on the particular machine in use, is automatically transferred into the loading position to receive the first tier of containers. The containers are directed in a single line into the machine. There they are arranged automatically into a pre-programmed load pattern and dropped onto the pallet. Successive tiers are reversed to lock rectangular containers together in the process. The completed module is removed from the load chamber onto a belt or steel-roller conveyor.

When slipsheets came into widespread use in the 1970s, it was necessary to convert the unitizers at many locations to build the loads on slipsheets. Machines called slipsheet inserters are placed adjacent to the pallet bank chamber, and vacuum cups are used to pick up and transfer a slipsheet onto the pallet prior to receiving the load. The slipsheet modules are then removed from the pallets and the pallets reused.

Another method of getting slipsheets under the loads involves the use of load transfer devices in which modules on a smooth-surface pallet are placed on a transfer device. A pusher plate is then used to push the load off the pallet and onto a slipsheet. The pallet is then recovered and stacked by the machine.

A lift-truck rotator attachment is used sometimes to install slipsheets in relatively low-volume operations. A slipsheet is manually placed on top of a pallet load. The palletload is then completely inverted so that the slipsheet is on the bottom. The pallet is then removed from the top and stacked.

FILM-WRAPPED UNITLOADS

The shrink wrapping of unitloads began in the 1960s. Shrink wrapping is said to have originated in Sweden for the shipment of bricks to construction sites. Palletloads of bricks were wrapped with plastic film to secure them onto the pallets. Someone noticed that the film shrunk tightly around the load when left in the sun, and the idea for shrink film was born.

The first applications of film wrapping involved placing a polyethylene bag over the module and then moving the module through a heat chamber to shrink the film tightly around the load. The thickness of the plastic film depended upon the weight of the load and required some experimentation. If the film is too light for a particular load, it could rip and tear in transit. The shrink-film method was effective but energy intensive, and thereby costly, particularly during the energy crisis of the early 1970s. This led to the development of the stretch-film method to reduce the amount of film required, and it eliminated the need for the heat chambers. With the stretch method, the film is simply stretched tightly around the load and the ends sealed. A number of machines have been developed for stretch-film applications. The methods of machine application include:

> The module is placed on a turntable and a band of film is attached to it. The turntable is then rotated horizontally to wrap the load.

> The module is positioned on a platform. An arm of a stretch-wrapper machine travels around the module as the stretch-film band spools out. Robot machines have also been used to travel around a module as the film is applied.

> The module moves on a conveyor to plow through a wide band of film. The film is stretched around the module, then cinched and tack sealed with an electric current.

The type of system that is selected will depend mainly upon the number of loads to be film wrapped per hour.

BULK CONTAINER HANDLING

If the project involves a new application of large bulk containers for the transport of bulk materials, it is important that the packaging specialist participate in the development of the specifications of the filling system. Whether it be a strap-hung bulk bag or a large paperboard carton, the interface of the module with the equipment is critical. There must be adequate space to set up the container and prepare it for filling. The containers are usually fitted with a plastic-bag liner for food ingredients. The materials may be diverted into the container from a conveyor belt or fed by gravity from an overhead hopper. Pumping systems are used to fill containers with bulk liquids, including aseptic bulk methods (see Chapter 18).

If the bulk container is to be shipped on a slipsheet, the slipsheet should be bonded to the bottom of the container at the time it is set up for filling. The slipsheet is placed into a squaring jig, and spots of glue are applied across the surface by either a brush or a pressurized dispenser. The empty bulk container is then set onto the slipsheet to bond it to the bottom. The K box type of container, which is described in Chapters 11 and 12, comes with the slipsheet permanently bonded to the bottom.

K boxes or containers that have adequately sealed bottoms that will not collapse or deform over fork tines can be handled by the take-it-or-leave-it method. The container is set up and filled on a pallet that contains open channels on the topside into which the fork tines can be inserted. The container can then be lifted and transported on the tines. A pusher device is used to hold the container in place as the tines are backed out. This kind of pusher attachment is about half the cost of attachments with the full push and pull functions. It is used for projects where limited shipping volume may not justify the purchase of the push-pull units.

Filling bulk containers with dry powder or granular materials may require a densification system depending upon the nature of the product and the rate of fill. This kind of system uses vibration to discharge air from around particles during the filling process and, thereby, compacts the material and improves the load density. Such a system is described in Case History #2, Chapter 12.

Whether the container is handled by straps, by pallet, or by palletless methods such as slipsheets, it must be capable of being moved from the filling station by mechanical means. This could be by a conveyor to move the module out of the filling area or with the use of mobile equipment.

If the module is to be shipped on a pallet it is very important to secure it to the pallet by straps. The width and strength of the strapping should be specified according to the weight of the module.

DELIVERY END MATERIALS-HANDLING EQUIPMENT

Machines that automatically depalletize small containers from unitloads are not common. Most breakdown operations today are manual. Where machines are used there are two basic types. The most common type makes use of a vacuum plate that is positioned over a tier of containers. A vacuum is drawn and the tier is lifted and transferred onto a conveyor belt to an unscramble machine that orients and lines up the containers into a single line. There may be a trend in the future to the use of robotic machines in which suction cups on the ends of robot arms are used to grasp and transfer containers from the unitload to a conveyor belt or other point of use. The design of the smaller container is most important to the operating efficiency of such machines. Poorly glued flaps or large gaps between the flaps can interfere with the suction process.

The discharge of flowable materials from large bulk containers is mostly mechanized in the applications that exist today. Liquid materials are pumped out either from the top by use of a wand or through valve fittings that are designed into the bottom of the container. Dry materials are removed by gravity by totally inverting the containers to allow discharge from the open top or by a built-in side gate at the base of the container. Nonfragile granules and powders can be removed by a vacuum process. The container is tilted to one corner, and a metal wand is installed in the corner. The wand is connected to a vacuum chamber by means of a high-pressure suction hose. When a vacuum is drawn, the material flows into the chamber, from where it is released to the point of use by means of a rotary air-lock valve or intermittent butterfly-type valve. Both the gravity and the vacuum systems are described in the case history on coffee powder in Chapter 12.

Many bulk module projects involve the conversion of existing small container systems to the bulk container methods. It is therefore important that the packaging specialist become knowledgeable about the filling and discharge systems associated with bulk modules.

8

Transportation

A great deal of change has taken place in transportation since unitized shipping was introduced in the 1950s. The first intermodal container service began in 1956, and it grew to become the dominant mode of ocean transport within the next 25 years. Intermodal trailer on flat car service (TOFC), grew steadily in the 1960s to become a major rail shipping mode in the United States. Trailers and semitrailers were used primarily for shorter hauls in the 1950s, but with the completion of the National Highway System, along with the growth of the TOFC intermodal system, trailers have become a common mode for unitized shipping.

The largest highway vehicle when unitized shipping was inaugurated in the food industry in the early 1960s, was the 40 ft semitrailer. Its length, 40 ft (12.2 m), and its width, 8 ft (2.44 m), had at the time been approved by most state highway authorities. The demand for longer and wider vehicles grew over the years. In

1982, the National Surface Transportation Assistance Act (P.L. 97-424) was passed. This law required all states to adopt new standards governing the length, width, and weight standards of vehicles operating on interstate highways (Figures 8.1 and 8.2). These were:

> Width: Limitation of 102 in. (2591 mm) on the interstate and national network highways with lanes 12 ft (3.66 m) or more wide.

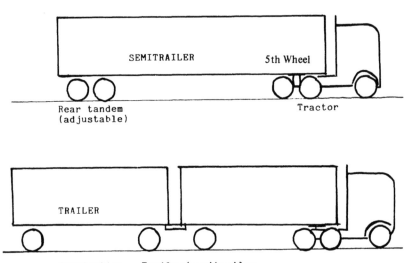

Combination - Trailer/semitrailer

Figure 8.1 U.S. NATIONAL SURFACE TRANSPORTATION ACT — 1982. OA width — 102 in. (2591 mm); OA length — semitrailers — 48 ft (14.63 m); OA length — trailer — 28 ft — (8.543 m); Maximum weight — 80,000 lbs (36,287 kg) gross
20,000 lbs (9072 kg) single axle
34,000 lbs (15,422 kg) tandem axle.
Some states allow operation of semitrailers up to 53 ft in length and triple trailer combinations up to 100 ft (30.5 m).

Contact the American Trucking Associations, Inc., Alexandria, Virginia, for current standards of all states.

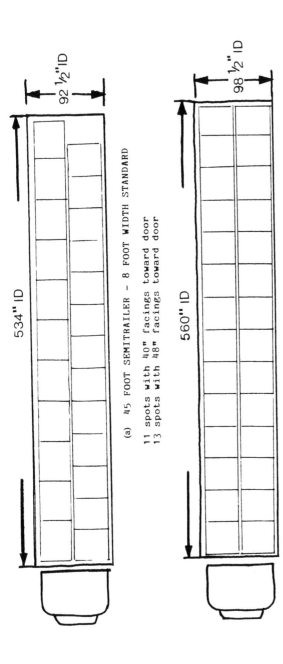

92 ½" ID

534" ID

(a) 45 FOOT SEMITRAILER – 8 FOOT WIDTH STANDARD

11 spots with 40" facings toward door
13 spots with 48" facings toward door

98 ½" ID

560" ID

(b) 48 foot semitrailer – with 8.5 foot width (new standard in 1982)

14 modules each side with 48" facings toward door

Figure 8.2 Standard 48×40 in. module configurations. (a) 45 ft semitrailer—8 ft width standard 11 spots with 40 in. facings toward door 13 spots with 48 in. facings toward door. (b) 48 ft semitrailer—with 8.5 ft width (new standard in 1982) 14 modules each side with 48 in. facings toward door.

Length: Semitrailers up to 48 ft (14.64 m) in a tractor/semi-trailer combination, or 28 ft (8.54 m) semitrailer in a tractor/semitrailer/trailer combination.

Weight: Limits of 20,000 lbs (9072 kg) single axle, 34,000 lbs (15422 kg) tandem axle, and maximum gross weight of 80,000 lbs (36,288 kg).

By 1989, 35 states allowed semitrailer lengths up to 53 ft (16.2 m) and an overall length with tractor up to 65 ft (19.8 m). The 80,000 lb gross weight of tractor and trailers with loads remained unchanged. Net payloads in the long semitrailers can be as high as 45,000 lb (20,412 kg). This can be accomplished by placing heavier modules forward in the trailer and repositioning the rear tandem to provide the right amount of counterweight to distribute the gross load weight over the tandem axle properly in order to stay within the legal limits. Computer programs have been developed to assist the planning of the load position and the setting of the tandem in order to achieve the highest payload possible.

Current information on size and weight limits can be obtained through the American Trucking Associations, Inc., 2200 Mill Road, Alexandria, VA 22314-4677.

The designer of transport modules should research information about the transportation equipment and systems through which the modules will be shipped. The inside dimensions are critical to sizing the modules to fit into efficient load configurations. The system flow chart should identify the kinds of transportation equipment to be used throughout the system. The design of bulk transport modules can unfavorably impact transportation costs in a number of ways:

Modules of light-density materials that are too tall to double stack can reduce net payloads.

The walls of modules that are designed with minimal material strength for the product shipped can bulge so much that the number of modules that can be fitted into the available space is limited.

Unitized loads of smaller containers that do not form into patterns compatible with the standard module dimensions can reduce payloads and cause damage in transit.

Transport modules that are undersized may require large amounts of dunnage to fill void spaces around them. The time and labor costs to install dunnage and to remove and dispose of it later, along with the cost of dunnage to begin with, can add substantially to total system costs.

The tare weight of packaging materials, if excessive, will increase the shipping cost per net ton of product shipped.

The height of a module may restrict transporting a double stack through standard vehicle doorways. This could require loading single modules at a time through the doorway and, thereby, double the time to load or unload the vehicle.

Although the length and width of semitrailers increased over the years, the 48 × 40 in. unitload continues to be the most common module length/width targeted dimensions. The amount of underhang and overhang tolerance, however, is not as critical for palletless modules, such as slipsheet loads, as it is for loads on pallets. The logistical infrastructure that includes warehouse racks and materials-handling equipment throughout the system will likely perpetuate the 48 × 40 in. standard for years to come.

To determine the dimensional tolerances that can be designed into the modules for specific applications, the designer should carefully research information on the inside dimensions of the transportation equipment to be used and establish the most efficient length, width, and height dimensions. This information can usually be obtained through a company's transportation or distribution department or directly from the carriers. If at all possible, the researcher should inspect the vehicles involved and take precise measurements.

The inside length, width, and height dimensions of the modules should allow reasonable operational tolerances. If, for example, the inside width is precisely 92 in. (234 cm), it would be impractical to design each module 46 in. (117 cm). At least 1 in. (25.4 mm) should be left on either side of the module. Likewise, if the doorway opening is 92 in. high, leave at least 3-4 in. (76–102 mm) above a tall module or a stack of two modules so that binding will not take place.

Special care should be taken in researching the inside dimensions and doorway clearances of refrigerated semitrailers, since the thickness of walls is related to the kind of insulation used. The following table that was compiled for a frozen goods project illustrates the differences that exist in equipment specifications.

	Carrier A	Carrier B	Carrier C
No. of semitrailers	370	380	350
Outside length	45 ft	42 ft	42 ft
Outside width	96 in.	96 in.	96 in.
Doorway ID height	101 in.	95 in.	97 in.
Inside length	44 ft	40 ft 7 in.	41 ft 6 in.
Inside width	91 in.	90 in.	89 in.
Cubic feet space	3200	2400	2339
Cubic meter space	90.6	68	66.2

The 48×40 in. standard module length/width dimensions could be used for any of the above. The limited inside width of the refrigerated units would require the modules to be positioned with the narrower sides all facing the doorway on the trailers of carriers B and C. The 91 in. inside width clearance of carrier A, while a tight fit, could allow a pinwheel configuration or combination of 48 in. and 40 in. facings to increase the number of modules per shipment.

Other information that should be researched for projects concerned with the shipment of frozen or refrigerated foods will include the refrigerant and air circulation system. Prior to unitized shipping, cases of frozen or refrigerated goods were hand stacked directly onto the decking of the rail cars or trailers. Channels in the flooring under the loads and wall ribbing provided air circulation around the load.

The efficiency of the refrigeration system requires air to circulate freely to and from the refrigeration units at the front of the vehicles. Residual heat from the products, or heat that transfers in through the floor, walls, and ceilings during the trip, is carried in the air stream to the unit, where it is removed and the cold air recirculated.

When the palletized shipping of frozen goods was introduced in the 1970s, the pallets provided a base under the loads through which air could circulate. As new refrigerated trailers were built in the 1980s, many were not equipped with adequate structural channels in the walls and flooring, since their designers assumed that pallets had become a standard means of shipping unitized frozen and refrigerated products. With the increased use of the slipsheet palletless shipping method, complaints of inadequate refrigeration began to be heard. Consequently, in 1985 an Interstate Carriers Conference called for a revision to the American Frozen Foods Association's manual of frozen foods shippers concerning modules on slipsheets. The revision would have required that all loads on slipsheets be transferred onto pallets for shipment.

Shippers who had already converted to the slipsheet method objected vigorously, since their reasons for using slipsheets in the first place was to eliminate the costs associated with pallets. Savings with the slipsheet method amounted to multi-millions of dollars annually for the large shippers. The sources of savings included:

> Initial cost of pallets compared with slipsheets. A pallet costs five or more times that of a disposable paperboard slipsheet.

> While pallets can be reused for several trips, the exchange process is burdensome and costly. It is difficult to maintain quality pallets in an exchange system. Few shippers make the needed repairs to pallets, and many pallets are lost in the system.

The pallets in a typical truckload add $1^{1}/_{2}$ tons of wood to
    keep refrigerated in transit.

Payloads are limited due to the space taken up by pallets and
    the tare weight they add to the gross load.

The position taken by shippers was that the carriers should provide
equipment with adequate refrigeration and air circulation systems
that would be consistent with current shipping practices.

The specifications that were identified to ensure adequate
circulation of air for the slipsheet loads included:

Insulation throughout the trailer to be adequate to minimize
    the transfer of heat through walls and flooring during
    transit

Ribbed walls or flat walls with a plenum chamber inside for
    air passage

T-bar floors, with channels at least $2^{3}/_{4}$ in. (7 cm) deep

Ceiling duct for distribution of returned cold air from the
    refrigerant unit to the doorway end of the vehicle

Freon 502 refrigerant

The issue was indicative of the lack of communication between the independent functional areas of logistical systems. The
carriers were apparently not aware of the reasons for the shippers
use of slipsheets. The shippers assumed that the trailer designers
would research the shipping methods in use and would design new
equipment accordingly.

This raises the question of whether shipping methods should
be adapted to the transportation equipment that is available, or
whether the transportation system should be changed to accommodate the shipping method. This will depend, of course, on the
scope of the project and the shipping volumes involved. In most
cases the shipping unit will be designed and dimensioned to most

efficiently utilize the equipment available. The case histories in Chapters 11 through 18 all involve packaging and shipping developments that were designed to be compatible with existing transportation equipment. Unitized shipping, on the other hand, would not have been possible without the vast changes in rail and truck equipment that were made in the 1960s to accommodate the new shipping method. Thousands of new DFB (damage-free bulkhead) rail cars that were especially designed for palletized shipping were put into service by the railroads at the time. They had heavy decking to bear the weight of lift trucks with loads, wall boards that could be extended to lock in palletloads, bulkheads to divide the car into compartments that better stabilized palletloads in transit, and wide doors to facilitate the entry of lift trucks. The trailers and semitrailers also required high ceilings and doorways through which stacks of two palletloads could be moved, as well as heavy floor decking that permitted the operation of lift trucks with loads inside the vehicle.

The introduction of the slipsheet method contributed to the proliferation of TOFC shipments. TOFC, or piggyback service, was initiated by the New Haven Railroad to accommodate intercity motor carriers in 1937. The method speeded rail service, since it eliminated the time and cost of switching rail cars from main yards to shipping docks. It did not really make a great deal of progress until 1965, when the National Piggyback Association was formed and carloadings topped one million a year for the first time. The method really took off in the 1970s, coincident with the implementation of slipsheet programs. The use of slipsheets made unitized shipping possible without the use of pallets under loads in the piggyback trailers. This made piggyback service acceptable to many shippers, who previously had avoided its use due to the cost of pallets and the impracticality of installing a pallet exchange system for piggyback shipments (Figure 8.4).

COFC (Container On Flat Car) shipments also expanded in the 1970s and 1980s. The minibridge concept was introduced in which trainloads of containers on flat cars were transported coast to coast.

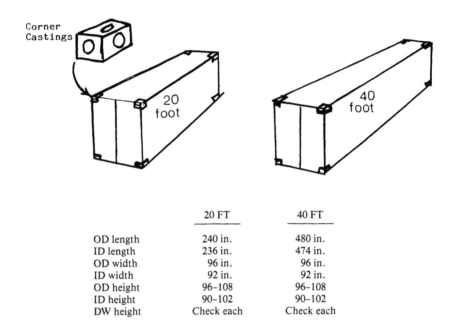

	20 FT	40 FT
OD length	240 in.	480 in.
ID length	236 in.	474 in.
OD width	96 in.	96 in.
ID width	92 in.	92 in.
OD height	96–108	96–108
ID height	90–102	90–102
DW height	Check each	Check each

LOAD CONFIGURATIONS WITH 48×40 in. TRANSPORT MODULES

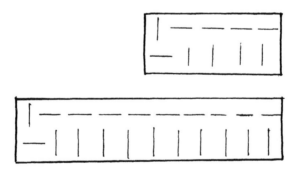

Figure 8.3 ISO dry cargo containers. The most common ocean containers are ISO 20's and 40's.

CONTAINERIZATION

Packaging specialists who will be designing transport modules for international shipments should be familiar with the standard ISO (International Standards Organization) containers. The overall and inside dimension of the containers are standardized, as are the floor channels in the refrigerated units. The only dimension necessary to check carefully is the doorway height and opening. The heights vary from 8 ft (244 cm) to 9 ft (275 cm).

Over 60% of the ocean cargo shipped today is containerized in standard 20 ft or 40 ft containers. In the early 1970s, the International Standards Organization adopted the uniform container standard that established international uniform specifications for ocean containers. Millions of the standard containers are in service today and are likely to be around for many years (Figure 8.3).

The intermodal ocean container concept revolutionized international shipping in a relatively short time. In 1956, Malcolm P. MacLean, owner of the Pan Atlantic Steamship Corporation, loaded 58 truck trailers, each 35 ft in length (10.7 m), and 8 ft (244 cm) height and width, aboard the Ideal-X, a converted World War II tanker, for a 4-day voyage from New York to Houston. The containers were hoisted aboard by means of a single gantry crane and latched to the top deck. In 1958, the Matson Navigation Corporation began the first container service on the West Coast, shipping 20 specially designed 24 ft long (7.3 m) containers on the deck of the Hawaiian Merchant, a cargo vessel, from San Francisco to Hawaii. In Hawaii the containers were transferred onto trailers for land transport to destinations on the island of Oahu.

The containerization experiments took place about the same time palletized shipping was developing in the United States. A key drawback to the unitized shipping of transport modules in ocean containers over the years has been the cost of pallets under the loads and the impracticality of recovering pallets for reuse in such a system. Consequently, even as we enter the 1990s, tremendous volumes of goods and materials are still hand stacked in and out of containers on international routes.

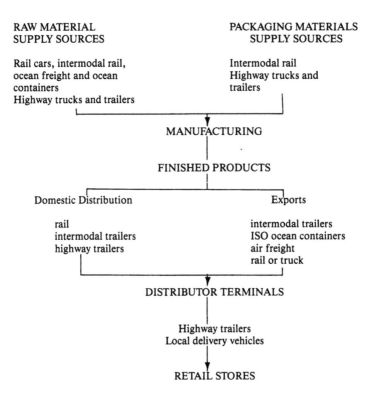

RAW MATERIAL   PACKAGING MATERIALS
SUPPLY SOURCES   SUPPLY SOURCES

Rail cars, intermodal rail,   Intermodal rail
ocean freight and ocean   Highway trucks and
containers   trailers
Highway trucks and trailers

                              MANUFACTURING

                          FINISHED PRODUCTS

        Domestic Distribution   Exports

            rail   intermodal trailers
            intermodal trailers   ISO ocean containers
            highway trailers   air freight
                                                rail or truck

                          DISTRIBUTOR TERMINALS

                          Highway trailers
                          Local delivery vehicles

                             RETAIL STORES

Figure 8.4 Transportation in a typical packaged goods logistical system.

The introduction of slipsheets has begun to make progress for international container shipments, but the lack of slipsheet expertise and handling equipment in many countries of the world has slowed conversion to the method.

Case History #2, Chapter 12, and Case History #9, Chapter 19, cover what is involved in the adaptation of the slipsheet method to international shipping operations.

Case History #8, Chapter 18, discusses the design of a new liquid packaging system to be compatible with the standard 20 ft ISO container.

9

Warehousing

The warehousing and storage of goods and materials is a necessary part of the logistical process. Logistics managers may reduce the number of warehouses and the amount of storage space required through the controlled deployment of inventory and the skillful management of production schedules. The need to warehouse, however, cannot be totally eliminated from logistical systems.

The purpose of this chapter is to familiarize the packaging specialist with the different kinds of warehouses and storage systems through which transport modules pass as the inbound materials flow to the manufacturing plants and the finished products flow to market.

The functional efficiency and the cost of most warehousing is dependent, to a large degree, on packaging design. It is important that the packaging designers be aware of the impact that the design of containers can have on warehousing efficiency and costs.

The kinds of buildings and storage systems will vary according to the fundamental needs for storage. Some key reasons for warehousing include:

To provide for the storage of agricultural products that are produced seasonally and held in bulk containers for distribution to manufacturers and packers for conversion to finished goods throughout the year

Examples: IQF frozen vegetables or fruits in large paperboard bins; fresh apples in wooden pallet boxes that are maintained in atmosphere-controlled storage rooms; dehydrated fruits and vegetables, nuts, powders, granules; aseptic packed tomato paste and concentrated juices

To provide for the storage of materials in process, such as cereals that are produced in large quantities, in order to minimize production costs

To provide for the storage of unitized loads or bulk containers at the point of manufacture, at regional market distribution locations, and at consolidation terminals

To store bulk bins or boxes of subassembly parts and pieces for manufactured goods

To provide storage space for empty containers and packaging materials to be used to pack raw materials, materials in process, and finished goods

To provide holding areas for goods awaiting quality assurance testing and staging areas for goods awaiting shipment

To provide storage and working space to depalletize unitloads, to consolidate and assemble many different products or pieces into unitloads for the delivery of raw and packaging materials to manufacturing locations, and for the shipment of finished goods to retail markets

Packing sheds along rail sidings at the edge of town, and multistoried brick warehouses in urban centers, provided the bulk of warehouse space in the first half of this century. Handling methods were simple and labor intensive. Two-wheel handcarts, platform carts, and rolling jacks with skids were used to transport products from docks to holding areas or into elevators for storage on higher floors. The throughput volume (the tonnage that could be moved into or out of storage in a day), was minimal and depended on the size of the crews available. Rail cars were held up at the shipping docks for days and even weeks at a time. Trucks and trailers often spent more time tied up at docks than on the highway.

Warehouse mechanization began to expand following the end of World War II as lift trucks and pallets were introduced. During the 25 years that followed, millions of square feet of single-story, high-roofed warehouse buildings were constructed in the industrial parks that sprouted up along mainline rail routes on the outskirts of cities and towns throughout the United States.

Typical postwar warehouse buildings provided raised docks, indoor rail sidings, wide-spaced structural columns, typically 40×25 ft (12.2×7.62 m), and ceilings 15–20 ft high (4.6 to 6.1 m) to accommodate the high stacking of palletloads by fork-lift trucks. As lift-truck technology progressed and multistage masts became available, the ceilings were pushed even higher in later years to allow for the higher stacking of pallets. The highest practical storage level for floor-operated fork-lift trucks is 20–24 ft (6.1–7.3 m). Above that, the added time to elevate the modules higher, along with the incremental investment in the higher reach lift trucks, must be weighed against the costs of buildings that have lower ceilings with more floor area in order to provide the equivalent storage capacity of the higher buildings.

The transition to the high-ceiling warehouse buildings is based on the economic premise that vertical storage space is more economical to construct and utilize per cubic foot than is horizontal space. That, however, impacts the guidelines for package design. Corrugated cases that are unitized on pallets and stacked 15–20 ft (4.6–6.1 m) high on warehouse floors must have sufficient top-to-bottom compression resistance designed into their walls to avoid being crushed under the stack-load pressure. The high stacking condition is further complicated if storage is in non-air-conditioned areas of warehouses during humid times of the year. The moisture can penetrate and weaken the walls of paperboard cases.

If packaging designers are not familiar with the kinds of warehouses and storage conditions through which their containers will pass, it is important that they obtain as complete information as possible for a first step in the design process. Tracing the step-by-step progress of products from the time of manufacture to final delivery by means of a simple flow chart will quickly build the knowledge needed. For example, assume the project involves the design of shipping containers for unitized shipping from a production plant that produces and ships packaged retail goods to a wholesale distributor terminal.

Manufacturing Plant

Product unitized on pallets and slipsheets. Stored in plant holding warehouse up to 2 weeks. Stacked four high in humidity-controlled storage area.
Product shipped two pallets high or three slipsheet loads high in rail cars to 15 regional warehouses. Stacked on slipsheets five high in non-humidity-controlled areas at 10 locations, and four high on pallets at five locations. Humidity above 50% in all locations from 2 to 9 months depending upon location. Maximum storage period at regional locations – 4 weeks.
Unitloads of different kinds of products are consolidated for shipment to wholesale terminal. Stacked two high on dock.
All products shipped two high on pallets in trailers to 60 wholesale terminals. Stored one high in racks at all but larger terminals, where stacked up to four high in non-air-conditioned areas. Humidity exceeds 50% at least 6 months of year in the larger terminals.

Such information does not necessarily mean the package design must satisfy the worst storage condition. On a systemwide cost basis, it could cost less to provide warehouse racking for products sensitive to crushing in high stacks, rather than increase packaging costs to provide greater compression resistance. The intent of the integrated packaging design approach is to minimize the overall system costs, and not simply to adapt packaging design to accommodate inefficient storage and shipping conditions.

Millions of acres of sprawling, single-story, high-ceiling warehouse buildings were constructed during the last half of the century and are likely to be around for some time. These buildings were relatively simple to design and construct. Their spacious interiors were adaptable to just about any storage configuration and all kinds of unitized products. If the need for warehousing diminished, the buildings were easily adaptable to light manufacturing operations. Structural methods and materials were varied according to the availability of materials and labor in different regions of the country, but certain functional design features are common to all. Floors and shipping docks were raised to accommodate the entry of lift trucks with unitloads into rail cars and trailers. The columns that support the roof are spaced to allow for the storage of standard unitload modules or to accommodate the installation of standard racking.

Despite the practical feature of single-story, high-ceiling warehouse buildings, there has been a trend to new and innovative kinds of warehouse structures since the mid 1970s. Computer-controlled systems, integrated with new kinds of automated and semiautomated materials-handling equipment, have given impetus to the construction of new kinds of warehouse structures that range from 30 to over 100 ft (9.1–30.5 m) in height. Warehouse structures can be classified into three basic types today (Figure 9.1): First, the conventional single-story type, as heretofore described, in which floor-operated lift trucks stack unitized loads and bulk containers 18–25 ft (5.5–7.6 m); secondly, mid-rise rack structures that are 30–40 ft high (9.1–12.2 m). Unitloads are moved to and from rack slots by very high mast stackers, some of which are fully automated and computer controlled, while others have operators aboard to manually direct the store and retrieve operations; thirdly, the very high rack structures of 40 to over 100 ft

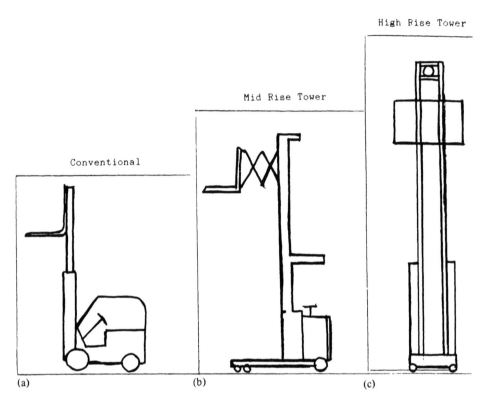

Figure 9.1 Warehouse lift vehicle types. (a) Conventional single story: storage height—18-25 ft (5.5-7.6 m), Floor stacked modules, racking optional, Rider driven counterbalanced or outrigger trucks; (b) Mid-rise Tower: 30-40 ft (9-12.2 m) totally racked, High-mast Outrigger lift vehicles, 5 to 8 rack levels, Mostly rider aboard (some are remotely controlled); (c) High-rise Tower 40 ft (12.2 m) to over 100 ft (30 m) Totally racked, Stacker cranes, 8 or more rack levels high, Mostly computer controlled.

(12.2 to over 30 m), most of which are fully automated and computer controlled.

Rack configurations in the mid- and high-rise systems vary according to the number of product codes stored. There are three general types of rack storage (Figure 9.2).

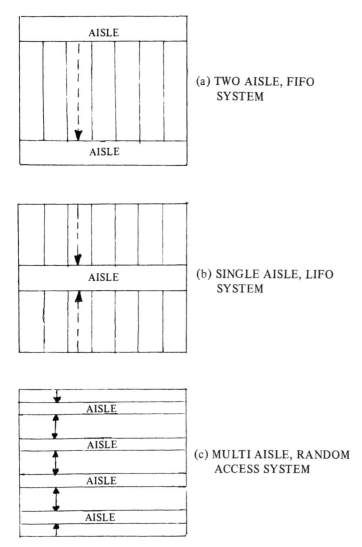

Figure 9.2 Rack configurations in mid- to high-rise warehouse towers. (a) Modules input on one aisle advanced through to aisle on opposite side. Either transfer carriers or rollers used to advance line as the front module is taken out. (b) Modules input into deep slots from a center aisle and taken out the same aisle. (c) A stacker crane is operated in each aisle.

FLOW-THROUGH TYPE RACKS

This type of rack system is also known as a live storage system. The unitload on a pallet is injected into a rack slot from an aisle by either a stacker crane or a high-lift vehicle. The unitload is advanced through the rack slot to the opposite end of the slot, where it is retrieved by another stacker or lift vehicle. There are different ways to advance the palletload through the slot. Sometimes this is done by means of gravity flow with roller conveyors on slightly pitched rack rails. Other methods include the use of wheeled pallets that are pushed along the rails or advance by gravity down a slight slope. Still another means of advancing the load is by use of a powered wheeled device, or dolly, that operates off the main stacker. This kind of device transports the pallet into the rack slot and mechanically lowers it onto the rails.

LIFO (LAST IN FIRST OUT) DEEP-SLOT STORAGE RACKS

This type of rack stores two or more modules deep on either side of an aisle. The modules are both stored and retrieved from the same side of the rack. The stacker, or elevating vehicle, may transport one or more modules at a time. These systems are found mostly at manufacturing locations where there are relatively few different products to store, and outbound shipments consist of truckload and carload quantities of the products.

RANDOM-ACCESS RACK SYSTEMS

These are racks that store modules just one deep on either side of an aisle in which the stacker or mobile lift vehicle operates. These systems, commonly called stacker crane systems, are the most common found in the world today. Typical rack configurations contain banks of back-to-back, one-module deep storage racks, which are separated by aisles.

Packaging designers should be aware of why the new and higher rack warehouses are appearing and how this might impact integrated packaging design in the future.

The transition from the thick-column, low-ceiling, multi-story urban warehouses to the single-story, high-ceiling warehouses following World War II came during a time when land was available for the development of industrial parks. Today many industrial parks are in congested areas, and the cost of land has risen sharply. Typically, for every square foot of space allocated to storage areas for floor stacks in the single-story warehouse, another square foot must be allocated for lateral aisles, cross aisles, and service areas for the materials-handling equipment. This, and the practical limit on the stacking height of conventional lift trucks of under 25 ft (7.62 m), increases the overall acreage required, to a point where many warehouse builders are looking to the mid-rise and high-rise warehouse structures as an alternative to reduce land costs.

Labor productivity is another consideration. Computers have been installed on lift trucks to help the lift-truck operators locate and assemble products more efficiently in many warehouses today. Still there is a limit to how far productivity can be increased with floor-operated lift trucks in a conventional single-story warehouse. The high-rack structures, in conjunction with new high-lift stacker vehicles, offer many opportunities for improving the productivity of manpower. Other key benefits of high-rack storage compared with the conventional warehouses include:

Reduced product damage: Compression of loads that is characteristic of floor stacking is eliminated. Also, the stacker vehicle paths are fixed, and maneuvering in tight storage slots is eliminated.

Gross space is better utilized: Aisles are narrower since the turning of floor-operated lift trucks within aisles is eliminated. Vertical storage configurations allow access to a greater number of unitload storage positions off a single cross aisle, thereby reducing the number of aisles required.

Improved control of goods stored: Each unitload has a discrete position in the rack structure. None are buried in stacks in deep storage slots. Stock rotation is more easily managed and inventory control is simpler. A greater number of unitload positions are accessible without disturbing other unitloads in storage.

Energy efficiency: The higher storage densities that are possible with high-rack systems make more efficient use of utilities such as lighting, heating, and air conditioning.

Financial benefits: Rack structures to which the walls and roof are affixed are recognized as equipment instead of buildings, and certain tax benefits are applicable. (Figure 9.3 – rack-supported buildings)

It is for these reasons that the number of high-rise and racked warehouses are likely to increase in the future. In many logistical systems the packaged goods and materials may pass through different kinds of warehouses. They may be stored in conventional single-story warehouses, where they will be high stacked, or they may be stored and retrieved in high-rack structures of different kinds, where they must fit into the rack openings.

If the packaging designer determines that the modules to be designed will pass through rack structures only in the system, it may be a relief to learn top-to-bottom compression strength of the modules is not that critical. There are, however, other packaging requirements of which packaging designers must be aware. Poor packaging is not tolerable in a rack system. Modules that bulge excessively or lean can get hung up or not fit at all into the rack openings. Modules that are inbound to storage are optically scanned, and if their dimensions exceed allowable tolerances of the system, the load is not allowed to pass through.

Whether conventional, mid-rise, or high-rise warehouses, the condition of the modules is critical to operating efficiencies. They must be dimensioned to fit easily into the storage slots. Most rack systems in the United States are designed to accommodate the standard 48×40 in. (1219×1016 mm) modules. Rack beams can be spaced to accommodate different heights of modules, but most accommodate modules up to 55 in. (1397 mm) tall, which includes the height of the pallet base. Modules must be properly prepared for storage as well as for shipment. Unstable modules, and those

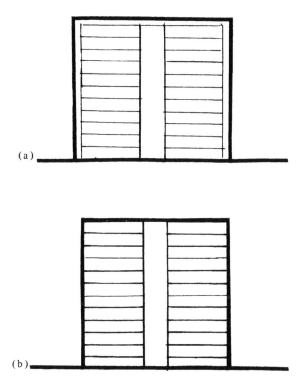

Figure 9.3 Mid-rise and high-rise structural design types. (a) Self-supported rack structure within a building. Racks may be taken out and replaced with racks in a different configuration. (b) Rack Supported — Sole Purpose Building. Used mostly for high-rise systems. Racks are strengthened to support walls and roof.

that bulge due to weak or collapsed cases, cannot be put through the warehouse systems. If the unitloads arrive on slipsheets, the condition of pull tabs is important. Torn crumpled tabs will make removal slow and difficult. If the unitloads arrive on pallets, broken and split boards and protruding nails will hamper handling through the warehouse.

The efficiency of breakbulk operations, in which unitloads are broken down or depalletized for assembly into consolidated unitloads, is sensitive to the condition of packaging and packaging design. Depalletizing, or breakbulk operations, are commonly combined with the order assembly process. The unitloads of products are set up in a line. Warehouse personnel, called order pickers, work along the line as they manually select the proper number of cases of each line item from the pallets and combine them into mixed unitloads for shipment to market.

The picking methods vary by distribution center. Some palletize the cases during the picking process. Others may load the picked cases onto wheeled carts, which are connected to draglines for towing to the shipping dock area. There they are disconnected from the line and moved into transport vehicles for hand loading the consolidated orders.

A few wholesale grocery distribution centers have attempted to mechanize the depalletizing operations and to combine them with an automated or semiautomated order sortation system. In those operations, the unitloads are depalletized layer by layer by means of a vacuum head plate. The plate is positioned over the top layer of the unitload, a vacuum is drawn, and the layer is transferred onto an unscrambling table that orients and dispatches the cases in a line on a conveyor belt to the order assembly area.

Packaging problems that have reduced the efficiency of the mechanized depalletizing process include: improperly glued case flaps that come loose, gaps left between the closed-top flaps, tacky palletizing adhesives used to stabilize the loads in transit, torn cases, loose wrappings or tapes, leaking containers, and unitload patterns that do not adapt well to mechanical unscrambling and orientation.

There are two basic types of automated and semiautomated order assembly systems. Both require the transport of the unitloads from the main warehouse storage areas to a breakbulk station where the individual cases are transferred onto conveyor belts. These are covered in the following subsections.

Batch Sortation Systems

Cases are coded by label, color, or audio input codes and transferred onto conveyors. A sensor system identifies the code in an analog memory control and tracks the progress of each case along the conveyor to a bank of chutes or bins. When the case reaches the proper bin or chute for a particular order, an electronic signal triggers a solenoid controlled gate and the case is automatically directed into the proper chute or bin.

SI Ordermatic System

This system is manufactured by S.I. Systems, Inc. of Easton, Penn. It is a system designed to automatically assemble up to 10,000 cases into retail store orders within an 8-hour shift. It is a unique system that, unfortunately, is limited in use due to the general condition of case-packaging in the food industry. The cases are directed from conveyor belts into a series of chutes 25 ft (7.62 m) in length, mounted on a 15° slope. The cases rest on plastic runners. A computer signal opens a gate at the bottom of the chute and meters out one or more cases from the chute onto a lateral conveyor belt that carries it to the unitizing station. Wider use of the system has been hampered by oversize or undersize cases, film-wrapped tray pack containers that do not slide freely down plastic runners, and weak unstable cases.

In both systems, the cases that are sorted by order are manually loaded into delivery vehicles in different ways. Some are conveyed from the chute system directly into the vehicles and floor loaded by hand. Some are hand palletized onto pallets, film over-wrapped, and staged for later loading by lift trucks. There are some operations in which the cases are transferred directly from the conveyors into rolling carts, and the carts are staged and are later pushed onto a transport vehicle by hand. The myriad of different sizes and shapes of cases or pieces has made the automation of unitizing operations for assembled orders impractical.

Marketing and sales demand dictate the size and shape of most shipping containers. There are limits on what the packaging specialist can do to standardize container sizes. There may be many opportunities to improve the efficiencies of warehousing and distribution center operations, however, if the packaging specialists have knowledge of the warehousing and storage systems in advance. As an initial step, packaging specialists should flow chart the systems and identify the warehousing operations through which the containers they design will pass.

Information on whether the goods will be stacked on conventional warehouse floors, injected into rack slots, or a combination of both should be obtained before attempting to design or make recommendations on the design of packaging. Wherever possible, packaging specialists should tour the total system and obtain details on rack slot dimensions and the mechanical systems that store, retrieve, convey, and handle the unitloads.

10

Research and Development for the Design of Transport Modules

There is an old saying that any fool can dig two post holes in a row, but digging three in a straight line requires planning ahead. The same reasoning applies to the design of transport modules for a total logistical system. The more complex the system, the greater the need for research and study of the entire system before attempting packaging design. Packing and shipping a product a short distance between just two points may not require a great deal of research and analysis in order to design a viable system. If the same product is shipped internationally, or over long-distance hauls, and must be handled at various intermediate stops in the process, information about the total system should be researched as a first step in the design process.

A good place to begin research is to determine if there are any specific rules and regulations governing the particular kind of product or shipping method that have been established by carrier or governmental agencies, or if there are any standard trade practices or guidelines for packaging, labeling, and shipping.

This does not necessarily mean that there could be restrictions of some kind that will prohibit the introduction of any new or innovative packaging system. It will, however, help to determine if transportation rate changes are indicated for the new system, and it will establish in advance the insurance coverage exposure in the event of an in-transit damage claim.

The importance of researching existing rules and regulations is particularly important when shippers convert from hand loading at the shipping docks to the unitized loading of transport modules. The hand loading of trailers is usually the responsibility of the carriers, and the labor cost is, therefore, reflected in the carriers' rates. Likewise, the accountability for the load count of pieces is borne by the carrier.

The loading and unloading of unitized modules, on the other hand, is usually done by the shipper or receiver using his own equipment and operators. In such cases, it is necessary to negotiate new rates that reflect the reduction in the carrier's costs to hand load and that transfers the responsibility for the piece count to the shipper. A list of trade and governmental agencies that issue rules and regulations for packaging, labeling, and the shipping of transport modules is included in the bibliography at the end of this book.

The amount of research that must be done to build useful knowledge about an existing physical system depends upon the complexity of the system and whether the project covers an entirely new product or is aimed at improving existing methods and costs.

In the early part of this century, scientific management pioneers such as Taylor, Gantt, Gilbreth, and others introduced a host of analytical methods and tools to study industrial activities. Motion and time studies, flow charts, statistical analysis, and many other systematic analytical techniques were used to obtain detailed information on the activities. From this knowledge base, certain general principles could be drawn, and the opportunities for methods improvement became apparent.

The application of sophisticated analytical techniques may not be needed to gain knowledge of system activities that influence packaging performance. Nevertheless, a systematic analytical approach to gaining information about the total system can speed the research process and help build a complete and accurate knowledge base from which decisions on packaging design can be made.

If the importance of the project justifies it, the research should include making visits to the shipping and receiving locations to observe the production, packing, and shipping methods, and also to become familiar with the activities along the logistical route. Flow charts should be put together in the process and photos taken for later study. If the particular project does not justify the time and expense for trips to the locations involved, the information may have to be obtained through telephone calls, correspondence, drawings, or photos. In either case, the packaging specialist should use a systematic study approach to ensure that complete information about the system is known.

An analytical tool that has been especially helpful for this kind of research is the flow chart or process chart. The amount of detail that is needed to put a flow chart together may vary by project. Assume, for example, that the project involves the need to improve packaging and shipping methods of fragile electronic subassembly automotive packages, such as the harness assemblies described in Chapter 16, case history #6. The project begins with a meeting of the persons at the receiving location who are concerned with product quality. Basic information about the system, such as the annual shipping volume and a sketchy summarization of existing handling and shipping methods, may be discussed. Samples of the product are exhibited. Each piece is $50 \times 10 \times 8$ in. $(1270 \times 254 \times 203$ mm). Four pieces are packed in a corrugated case, and eight cases are unitized on a wooden pallet. They are shipped in truckload lots over a 900 mile route from the producing plant to the user location. The incident of damage is very high, and the objective is to find a way to eliminate or reduce the damage. The most common damages found are hairline cracks to the high-density plastic housings of the wiring assembly.

A decision may be made to study the possibility of redesigning the housing's mold or to use a more durable plastic material to withstand the shocks and vibrations the pieces take during highway transit. In this example, we will assume the decision has been made to ship a number of the damaged pieces to the packaging laboratory for analysis to find ways to improve the packaging to reduce the incidence of damage.

The traditional approach would probably be to study ways to strengthen the packaging or to stabilize the piece with cushioning materials. Hand-built prototypes would be put through a series of dynamic load tests. Eventually, a decision on a stronger and improved packaging method may be forthcoming. It would very likely result in an increase in the packaging costs per piece shipped. However, since packaging changes are indicated, it would be a good time to introduce a thorough study of the total logistical process. The management objective would be broadened to determine if opportunities exist for overall cost reduction in the logistical process, in addition to solving the immediate damage control problem. Management may agree and assign the project to the packaging area, since packaging is the basic function that influences costs throughout the system.

It is now up to the packaging specialist to carry out the research necessary. The best means of obtaining the needed information is to visit the production and user plant operations and to flow chart the sequence of activities in the handling, shipping, and receiving operations.

Field trips to the key locations will not only provide the factual information needed, but will afford an opportunity to gain the input of workers and management people at those locations. First-hand observations may also provide the creative packaging designer with ideas for innovative methods. Photos and slides should be made to supplement the flow chart. They can be invaluable in later analysis and in communication of the opportunities for design changes to others.

Figure 10.1 illustrates a simple flow chart that could be used to ensure all pertinent details are covered in the analysis. The technique is to begin with the production process and to trace the

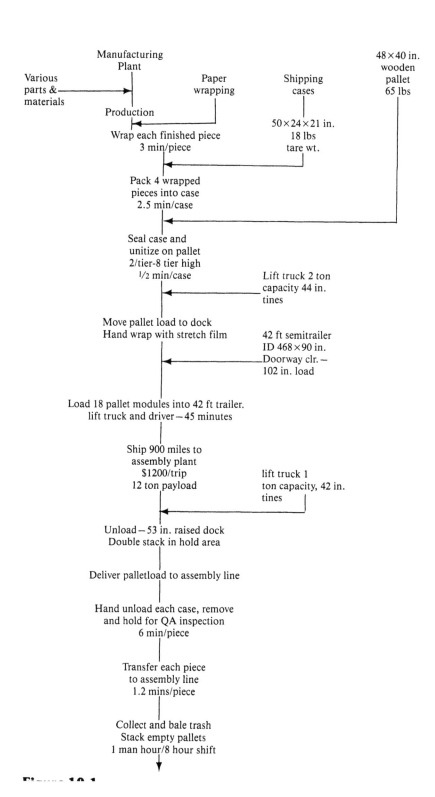

Various
parts &——————►
materials

Manufacturing
Plant

Production

Paper
wrapping

Shipping
cases

48×40 in.
wooden
pallet
65 lbs

Wrap each finished piece
3 min/piece

50×24×21 in.
18 lbs
tare wt.

Pack 4 wrapped
pieces into case
2.5 min/case

Seal case and
unitize on pallet
2/tier-8 tier high
½ min/case

Lift truck 2 ton
capacity 44 in.
tines

Move pallet load to dock
Hand wrap with stretch film

42 ft semitrailer
ID 468×90 in.
Doorway clr. −
102 in. load

Load 18 pallet modules into 42 ft trailer.
lift truck and driver − 45 minutes

Ship 900 miles to
assembly plant
$1200/trip
12 ton payload

lift truck 1
ton capacity, 42 in.
tines

Unload − 53 in. raised dock
Double stack in hold area

Deliver palletload to assembly line

Hand unload each case, remove
and hold for QA inspection
6 min/piece

Transfer each piece
to assembly line
1.2 mins/piece

Collect and bale trash
Stack empty pallets
1 man hour/8 hour shift

Figure 10.1

steps in sequence in a flow pattern. Materials, handling devices, and the equipment used are listed horizontally across the top of the chart. Lines extending from these listed items indicate where they are brought into the process. Pertinent factual data that can impact packaging design are jotted on the appropriate lines in the flow chart. Any other bits of information that may be picked up are also jotted down as the flow process is traced. Sometimes the suggestions of workers are worth noting at a particular step in the procedure. Information on productivities and the costs of any step in the procedure may also be noted on the chart. The chart is strictly for the benefit of the observer and not intended to be used for presentations or reports.

When the flow chart is completed and all pertinent details collected, the next step is to summarize the information into functional categories for final evaluation. Figure 10.2 illustrates the summarization of information for this example. Pertinent details are listed under the headings of packaging, transport module, handling time, and shipping. Other headings may be used to provide greater detail on specific functions, such as the intermediate transfer operations along the route if they are part of the process. There are no mandatory rules for such charts. It is up to the analyst to determine how to summarize the information for the convenience of review. What is most important is the systematic documentation of all pertinent information.

Next comes the assessment of the information collected and the identification of opportunities for improved system efficiency through the packaging design process. In reviewing the flow chart and the summary of factual information in this example, certain possibilities for improving the methods become apparent. The key areas indicated for improvement are:

Reduce Transportation Costs

The existing pack method allows only 336 pieces per truck-load, which gives a very light highway trailer load weight that is under 10 tons. Up to 20 tons can be shipped in a standard highway trailer without exceeding the road weight limits. Large savings in transportation costs could be realized if the number of pieces per

PACKAGING

Shipping case size -- 50×21×24 in.
                                                                  14.5 cu ft–0.41 cu m
pieces/case --- 4
Net weight per piece -- 35 lbs (15.9 kg)
Tare weight --- 18 lbs (8.1 kg) per case
Stacking height --- 14 feet (normal conditions)
Style case -- tube and caps
Material -- double wall corrugated
Pallet type --- wooden, 48×40×5 in.
Pallet tare weight -- 65 lbs
Stretch-film overwrap --- 2 mil PVC

TRANSPORT MODULE

Cases per pallet tier --- 2
tier per pallet --- 4
Overall height – pallet and load ------------------------------ 102 in.

HANDLING TIME

Man minutes to wrap each piece -------------------------------- 2

Man minutes to set up case and pack
4 pieces in it -- 5.5

Man minutes to unitize 8 cases onto pallet and
stretch film overwrap --- 6

SHIPPING

Transportation vehicle -- 42 ft semitrailer
Doorway clearance --- 102 in. (259 cm)
ID dimensions --- length, 498 in. (12.6 m)
                                                                 width, 90 in. (229 cm)
Total # pallets/truckload ------------------------------------- 21
Configuration --- pinwheel (10 deep one side
                                                                 11 deep other)
Cases/truckload --- 84
Pieces/truckload -- 336
Packaging tare weight --- 6048 lbs (2743 kg)
Pallet weight (21 pallets) ------------------------------------ 1365 lbs (619 kg)
Net weight – (336 pieces) ------------------------------------- 11,760 lbs (5334 kg)
Gross load weight --- 19,173 lbs (8697 kg)
Freight charge/trip --- $1500.00
Freight charge/piece shipped ---------------------------------- $4.46

Figure 10.2 Summary of pertinent information.

shipment could be substantially increased. The tare weight of packaging and pallets amounts to over one-third of the gross load weight. The reduction of tare weight, along with an increase in the number of pieces per shipment, would further reduce the transportation cost per piece.

Reduce Handling Costs at the Point of Origin

The setting up of the cases plus the packing of just four pieces in each case and unitizing the cases onto pallets is labor intensive. Hand wrapping the cases with stretch film to secure the transport module together adds further to the costs. A larger case or bulk container should be considered.

Reduce Handling Costs at the Receiving Location

As on the production end, the number of cases to handle and open is labor intensive. Larger containers make sense.

Reduce the Amount of Waste Packaging

For the number of pieces involved, there is a great deal of waste packaging material, as well as pallets, to collect for trash disposal. There are over 3 tons of waste to dispose for every truckload of 336 pieces.

Eliminate the Source of Damage in Transit

The configuration of the transport module (two cases/tier and four tiers per module) results in an overhang of 2 in. (5.1 cm) in both length and width pallet dimensions. This condition could contribute to the damage control problem. The stacking of pieces two high within cases stacked four high could build up load pressures that contribute further to the damage experienced.

With the opportunities for improved costs and productivities known, the methods of improvement become readily apparent.

Develop one large bulk container that will have the capacity for an entire pallet load of the existing cases.

Pack the harnesses tighter together in the bulk container to increase pack density.

Design the large container so that the pieces can be packed directly into them at the production line and removed directly from them for immediate delivery at the point of use.

Develop an integrated pallet/container system to avoid handling separate inventories of containers and pallets.

Consider the design of a durable container that can be economically collapsed and returned to the supply source in order to eliminate trash disposal costs associated with the existing method.

Develop a means to reduce load pressure on any individual piece within the module.

The packaging specialist is now ready to work out the design options. Elements of the above example were taken from the case history described in Chapter 16 on the harness box. The innovative vertical layering packaging system that was developed for that project demonstrated well the benefits of systematic analytical research as the first step in the packaging design process. The research and study identified the design objectives. The skill and experience of the particular packaging design consultant were instrumental in finalizing the design features needed.

The packaging designer developed a packing system in which each piece would be suspended on a strap hanging from a bar across the top of the box. In that way the internal load pressures that resulted in damage were eliminated. He then installed the strap system in a large box that contained a removable front panel for access to load and unload the pieces. To eliminate the trash disposal problems, the box was designed to collapse into a compact package for the return haul, and it was built to be durable enough to last for a number of trips.

The building of a number of prototypes may be necessary before the ultimate design emerges. In the case history example, the first prototype was 89 in. (2261 mm) tall, 60 in. (1524 mm) long and 45 in. wide (1143 mm). The main purpose of the first prototype was to test the innovative strap suspension method. The dimensions of the first prototype built were based upon the objective of getting as large a box as possible to minimize the number to be handled throughout the system.

As it turned out, the strap system worked well, but the size of the big container was unwieldy and difficult to load and unload at the production lines. A second prototype was built that measured 58 in. (1473 mm) long, 45 in. (1143 mm) wide and 50 in. (1270 mm) high. The containers each contained 45 pieces. They could be stacked four or five high in warehouses and shipped two high in semitrailers.

Much can be learned about the performance of a transport module of this kind by putting the prototypes through a series of dynamic load tests in the laboratory. The lab selected for the tests should have equipment sized and rated for bulk containers. Three basic tests are generally used:

> High-amplitude, low-frequency vibration test to simulate the forces experienced in over-road hauls. One or two modules are placed onto a vibrator deck and vibrated at 200–250 cpm for an hour or more.

> Incline, shock/impact test in which the module is placed on a dolly, pulled up an incline, and released to crash against a wall.

> Top-to-bottom compression resistance test in which the module is placed into a machine that lowers a plate across its top and gradually increases pressure until the failure point is reached. As a guideline, large bulk containers should be designed to withstand a force at least four times that of the maximum load placed on it in a warehouse stack.

The lab tests may be expanded to include the stackability of the containers under varying humidity and temperature conditions, as well as with the materials-handling equipment to be used in the system.

Whatever tests are run, it should be kept in mind that tests conducted in the laboratory give only an indication of the performance of the module under actual shipping conditions. Obvious flaws and wear-and-tear points can be corrected and materials specifications firmed up from the results of the lab tests. The module is not ready for final production, however, until it has been put through one or more shipping tests.

The input from workers who will handle the box in the production and logistical operations is an excellent source for fine-tuning modifications. The field tests can also provide information on the training needs of the workers. For example, the process of latching the pieces onto the straps in the vertical layering system requires a knowledge of the techniques and procedure to pick up a piece and orient it into the proper position for latching onto the straps. It should not be taken for granted that simple procedures will be obvious to the worker and that no instruction is required.

The final step is the implementation of the system. It is advisable to continue to monitor the project to ensure that the system is performing as planned and that it is producing the cost savings and the functional benefits expected.

The foregoing example had its origin in the need to find ways to improve an existing packaging system. If the project involves a new product for which no past experience is available, the preliminary research would be directed toward getting as much factual information as possible about the product characteristics and the distribution channels through which the product will flow.

As a sample project, assume a meeting is called to discuss the production of a new breakfast cereal product. Basic information about the product is discussed, such as predicted volumes, sources of supply for the ingredients, and marketing plans for distributiion of the finished product. The cereal contains a mixture of two kinds of granular materials that will be obtained through two suppliers in different parts of the country. During market tests, when relatively small supplies of the product were produced, the ingredients were obtained in small containers and bags that were manually handled. Now that the product will be going into volume production, the possibilities of a bulk container system should be studied.

The assignment for the packaging area is to research and make recommendations on a bulk container system for packing and shipping the ingredients from the supplier sources to the central plant location. The first step is the preliminary research to build knowledge on the total system. Since it will be a new system, the chart can be made without the need for field trips at this time.

Information on each ingredient, such as density, the effect of different temperatures and humidity on the product, its susceptibility to infestation, and other information that might affect the container specifications should be obtained.

For the purpose of illustration, let us assume the density of one of the ingredients is 10 lb/cu ft and the other is 30 lb/cu ft (160 kg and 480 kg per cu m).

Information on transportation equipment that will be available to haul truckloads of each ingredient from supplier locations should be noted on the flow chart. In this case we will assume that trailers 48 ft (14.63 m) in length are available. A trailer that size can contain 28 standard 48×40 in. module positions.

Further flow charting determines that upon arrival at the central plant, the ingredients will be taken to a storage area and held for quality assurance tests. Lift trucks with fork tines or push-pull attachments are available for that purpose. The fork trucks have a carrying capacity of 4000 lbs (1814 kg) at 24 in. (610 mm) load center when equipped with standard fork tines. When equipped with the push-pull attachments, the carrying capacity at the same load center is reduced to 3200 lbs (1452 kg)

Notations should be made on the flow chart concerning access for lift trucks to the discharge area and the amount of space available for the installation of bulk container discharge equipment such as a tilt table.

During the flow charting process, the suggestions and objectives of each functional area in the process should be noted. The objectives are listed for the buyer end and then followed up with the suppliers to determine if they can adapt their operations to accommodate the buyer's objectives.

Examples of functional objectives:

Central Plant Production Manager

Standard size box for both materials.

Size of bulk boxes should be as large as possible in order to minimize the number to handle.

Discharge rates should be the equivalent of four pallet loads or 200 bags, 10,000 lbs (4536 kg) per hour for the heavier product and 3000 lb (1361 kg) per hour for the lighter product.

Transportation

Minimize lightweight load penalties.

Modules to be sized to maximize the use of cube inside 48 ft semitrailers for light products and to distribute load weight of the heavier product uniformly throughout the trailer.

Eliminate voids around the modules that would require the use of dunnage.

Materials Handling

Use of paperboard slipsheets for shipping to avoid the cost, collection, and return of pallets.

Gross weight per module not to exceed 1000 lb (454 kg), to be compatible with existing materials-handling equipment.

When the flow analysis is complete, and as much information as possible about the project is documented, the next step is to determine the optimum size and dimensions of the container. The range of densities will make it difficult to obtain a standard-size box and to satisfy all of the functional area objectives. Compromises may be necessary in order to get the most effective size box to benefit the total system.

The length and width dimensions of the bulk box must be modular to the inside dimensions of the transport vehicles. The standard 48×40 in. (1219×1016 mm) pallet size would be applicable. The approach, therefore, will be to design a bulk container with a 48×40 in. base and to select a height that will result in a container with an inside cubic capacity that can be cost effective for either the light- or heavy-density ingredients.

It would be simpler to select the height to suit each ingredient, but the objective is a single-size box. The lightest weight ingredient, 10 lb/cu ft, would benefit from the tallest box possible that could fit through the doorway of the vehicle or 99 in. tall (2515 mm). A container that tall, however, would be entirely unsatisfactory for a number of reasons. First, it would be unstable, and its pliable paperboard walls would be subject to severe bulge. If filled to the top, it would contain 1070 lb (485 kg) of the light-density ingredient and 3210 lb (1456 kg) of the heavier material. The heavier loads would limit the number that could be shipped, since the gross load would far exceed highway load limits of 20 tons.

A compromise would be a container short enough to stack two high in the transport vehicle. Considering the doorway opening and allowing at least a 3 in. operating tolerance, the height of the shorter container would be 45 in. (1143 mm). Such a container would provide approximately 45 cu ft (1.27 cu m) inside capacity. Net weights of the lightest and the heaviest ingredients in the 45 cu ft box would be:

@10 lb/cu ft — 450 lb (204 kg)

@30 lb/cu ft — 1350 lb (612 kg)

The 48 ft semitrailers to be used allow a configuration of 28 floor positions for the 48×40 in. standard. The containers with light ingredients would, therefore, be shipped two high in the trailer, or 56 in all, and the heavier product would be shipped one high, or 28 per shipment. The tare weight of each box based on the total square footage and the thickness of the corrugated paperboard used can be estimated to be about 75 lb (34 kg). The approximate truckload gross weight to be carried can then be derived as follows:

lightest ingredient — (75+450)×56=29,400 lb (13335 kg)

heaviest ingredient — (75+1350)×28=39,900 lb (18098 kg)

The truckload weights in some cases can be increased by deaeration of the material during filling. A vibration method that had been used successfully for deaeration of powders and granules is described in Chapter 12. The amount of increased density possible is related to the fill rate. Unless the fill rate is very slow, most granular materials can be increased 15 to 40% by the vibration method. Increased pack density will reduce the number of containers and the number of truckload shipments required for the same volume shipped. This possibility should be communicated to the buyer of the ingredients and may involve visits to the suppliers' plants to discuss details of the vibration method.

Since the heavier containers are shipped one high only, they open another possibility for increasing the payloads per shipment, that is, to stack the additional modules two high in the front end of the semitrailer (over the fifth wheel). In this example, two additional containers could be added to increase the gross load weight to 42,750 lb (19,391 kg). This kind of loading requires the side walls of the trailer to be fitted with linear metal strips for securing either web straps or crossbars across the back of the unbraced containers to prevent them from sliding about in transit.

There could be other factors that will influence the size of the container, such as the batching of ingredients. These should be noted in the objectives and considered in the analysis. In this example we will assume that batch size is not a factor to consider.

The project is now ready for the construction and testing of the prototype containers.

TRAINING AND FOLLOW-UP

Many compromises normally have to be made during the demonstration of new methods for shipping tests. The objective of shipping tests is not only to gain information on the kinds of modifications to the packaging design or handling equipment that may be needed, but to build confidence in the new method so that the people who will be using it will feel comfortable that they can make it work. The importance of supervision and instruction in even the simplest of methods can be important.

11

Case History 1: Bulk Raisin Modules

The successful design and development of a containerized bulk module developed in 1980 for raisins provides an excellent example of integrated bulk packaging design.

Raisins are used in a variety of breakfast cereal products. In the United States, most are produced in the central valley of California and shipped from there to cereal packers throughout the country. Prior to 1980, the raisins were shipped in 30 lb (13.6 kg) corrugated cases. The cases were hand stacked into RBL type (air-conditioned) boxcars for the long cross-country hauls. As production volume grew, the costs of handling these small cases led to experimentation with mechanical handling methods with large corrugated bins on pallets. Polyethylene liner bags were installed in the bins, filled with raisins, and heat sealed at the top.

The bins were dimensioned 48×40 in. (1219×1016 mm) to be compatible with the standard food industry wooden pallet. The bin height, however, was restricted to 20 in. (508 mm), since larger capacity bins caused the raisins to agglomerate into large clusters that could not be broken up easily to flow onto packing line conveyors. The 22.2 cu ft (0.63 cu m) bins held just 1000 lbs (454 kg) of raisins. In addition to providing mechanical handling capability, the rigid wooden pallets bridged over the tops of the large bins and prevented the top bins in a stack from caving into the lower ones. The standard pallets adapted to a 28 pallet stack configuration in the rail cars. Palletized bins were shippable in three-high stacks for a total of 84 bins or 84,000 net lb (38,102 kg) per carload.

The paperboard bin-on-pallet system was a significant improvement over the 30 lb case methods. Each carload of 84 bins eliminated the manual handling of 2800 thirty pound cases. Discharge of raisins from the bins was done by inverting the bins on mechanical tilt tables to allow the raisins to flow out by gravity. The empty liner bags were discarded as trash, but the corrugated bins were collapsed flat, strapped together into piles, and returned to raisin suppliers for reuse. Pallets in RBL rail cars were not part of the industry exchange program, and their return for reuse was cost prohibitive. Some were sold for scrap lumber, while others were shipped to nearby plants or warehouses for reuse.

With slipsheets making inroads in the 1980s, and the need for surplus pallets greatly reduced, the next step was to adapt the bulk bins to slipsheet handling. Without the rigid pallet base, a key packaging design concern was strengthening the bins adequately to prevent their collapse in a stack of two or more. Another needed improvement, based on experience with the IQF bins, would be a way to fasten the slipsheets to the bottom flaps so that the bins could not slide over the tab scorelines and hamper unloading by push-pull attachments. The type and strength of the slipsheets selected were critical to the success of the program. Heavy-duty slipsheets, .095 caliper thickness, coated with polyethylene on both sides, worked well. The sheets were glued onto both top and bottom flaps of each bin with a special breakaway-type, cold-set glue. This locked the bins securely onto the sheets to prevent sliding about in shipment. The friction surfaces of the polyethylene-coated sheets between bins in a stack contributed further to load stability.

The procedure was to position a pallet on the roller conveyor line to the filling hopper. A slipsheet was then placed onto the pallet and a squirt bottle used to dispense spots of glue onto its surface. An empty bin with bottom flaps taped closed was set onto the slipsheet to bond it to the flaps. Next, a polyethylene liner bag was installed into the bin, and it was rolled onto a scale under the fill hopper, where 1000 lbs of raisins were metered into it. The liner bag was heat sealed at the top and the top flaps of the bin taped down. Glue was then dispensed to the top flaps surface and a second slipsheet applied over it. When a filled and sealed bin reached the end of the conveyor line, a lift truck with a push-pull attachment removed it from the pallet and transported it to a staging area to await shipment. The empty pallet was taken back to the front of the line for reuse.

An important design change for conversion from pallets to the slipsheet method was the dividing of each bin into two interior cells. An upright center piece of board provided the compression resistance needed to ensure that top loads would not cave into the tops of lower bins in a stack. It was also found necessary to strengthen the bins at the corners. The modifications increased the price of the bins approximately 10%. Packaging costs per net ton shipped, however, were substantially less for the slipsheet method, since the cost of pallets was eliminated. Also, without the pallet taking up space under the bin, it was found possible to increase the bin height 3 in. (76 mm) and pack an additional 100 lbs, (45.4 kg) of raisins into each bin to further reduce the packaging cost per net ton shipped.

A major transportation cost benefit was found in the slipsheet method, since with the pallets removed the bins could be shipped four high in the rail cars for a total of 112 bins per carload. This method, along with the added 100 lbs per bin, increased carload net payloads from 84,000 lbs (38,102 kg) to 123,200 lbs (55,883 kg). The greatly improved payloads for the air-conditioned RBL rail cars resulted in substantial transportation cost savings.

During the implementation of the project, it was observed that the bins could be more effectively integrated into the system if certain design changes could be made. Included were the following:

Design the slipsheet as an integral part of the bin. This would eliminate the need to glue slipsheets onto bottom flaps at the supplier end and the need to rip them off in order to collapse boxes on the user end. It would further avoid the need to purchase slipsheets and bins separately and to carry separate inventories of each at the packing plants.

Eliminate the top and bottom taping down of flaps. Tapes had to be cut to open the bins and to collapse empty ones. Bits of tape that ripped off the bins became a disposal problem.

Design a bin that will collapse into its 48×40 in. base. The empty RSC style bins were 88 in. (2235 mm) long when collapsed and required two people to set them up.

Design bins with a four to one top-to-bottom assurance factor. Bins were stacked up to six high in warehouses and staging areas of shipping docks, and shipped four high. Since the slipsheet bin was relatively new, no previous guideline for top-to-bottom compression resistance existed. With a four to one assurance rating, the bottom bin in a stack of two or more must be designed to withstand a top-load pressure equal to four times the weight of the normal load to be placed on it in a stack. If bins were stacked six high in warehouses, the gross weight on the bottom bin in the stack is approximately 5750 lb, (2608 kg). The top-to-bottom compression specification would be 23,000 lb (10,432 kg)

A number of design alternatives were studied and prototypes built and tested over a period of a year. The most successful of these was the KR box, the *K* standing for the initial of the designer's name, Kupersmit, and the *R* for raisins (Figures 11.1 and11.2).

The KR box was a radical new design. Unlike other large boxes, it could not be manufactured totally on a single production line. It required a secondary manufacturing process to assemble the parts. The incremental cost of the secondary process resulted in a 15% to 20% higher price compared with RSC-style bins that

OD dimensions set up $-48\times40\times23$ ins. $(122\times102\times58$ cm)
Tare weight 50 lbs. (22.7 kg)

(a) Cap $-48\times40\times5$ ins. $(122\times102\times13$ cm)
     $-$ 1100 lb test triple wall
     $-$ 90 lb liners outside
     $-$ 69 lb liners inside
     $-$ 33 lb semi-chem mediums

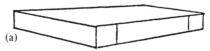

(b) Two load bearing inserts
     $-$ collapsible into base
     $-$ 1100 lb test triple wall
     $-$ 90 lb liners
     $-$ 33 lb Hydro Chem
       or wax-impreg-
       nated mediums
     $-$ nonreversible score
       lines
     $-$ compression test
       rating $-$ 18,500 lbs
       (8392 kg)

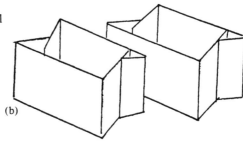

(c) Collar
     $-$ 600 lb test double wall
     $-$ 90 lb liner outside
     $-$ 69 lb liner inside
     $-$ 33 lb semi-chem
       medium
     $-$ RSC 4 flap bottom
     $-$ .090 PE coated slipsheet
       with 3 in (76mm) tabs
       on one long and one
       short side.

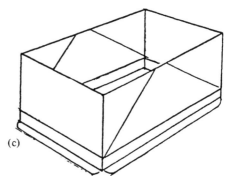

PE liner bags: $38\times24\times72$ ins. $(965\times610\times1829$mm) each cell. Mil thickness and MVT factors depend on product type. (Option $-$ single bag to drape over the center divider walls of the inserts.)

Figure 11.1 KR Type box $-$ 40 lbs/cu ft or more capacity.

Figure 11.2   KR slipsheet box for bulk raisins.

could be manufactured on available production line equipment. The incremental cost, however, was justified by the system savings that the functional characteristics of the unique box made possible. It demonstrated that the primary objective in integrated bulk container design is more than simply achieving the lowest possible container price.

In order to collapse the bin within its 48×40 in. base, the slipsheet had to be permanently bonded onto the bottom flaps. Diagonal scorelines were designed into the walls so the bins would collapse like a bellows onto the base. Since the scorelines would weaken the walls and preclude achieving the required top-to-bottom compression specification, the bin was designed in two sections. A relatively light, double-wall board was used for the exterior tube to which the slipsheet was bonded. The tube provided a protective housing that adapted well to scorelines, but it was not intended to contribute top-to-bottom compression resistance. Load-bearing specifications were accomplished through the use of two inner cells that were designed of heavy-duty, 1100 lb test, triple-wall board. Vertical scorelines on the ends of each inner cell made it possible to collapse the cells flat for convenient storage inside the base of the exterior tube.

Setting up a bin of this kind was easily done by one person in a matter of seconds. The top cap is removed and the outer walls pulled upright. The cells are then pulled into position and snapped into place.

Compared to the RSC-style box with glued-on slipsheets, the KR boxes substantially improved the productivities of the crews at both the filling and discharge operations. An unexpected benefit was found in that this kind of box reduced the costs of packaging over a period of time. The outer tubes and slipsheets, which are most vulnerable to wear and tear, have to be replaced periodically but the inner cells can be reused many times.

The K box concept was eventually used for a variety of raw materials. Modifications were made to the outer tubes for boxes of lighter density materials in which the box height exceeded the width. For these boxes the scorelines were designed to collapse the tube onto the base in a pinwheel fashion. A major application of the modified box, the KC box, has been used extensively for shipment of relatively lightweight soluble coffee powders. (See Chapter 12 on the case history of the development of a coffee-powder transport tote system.)

12

Case History 2: Coffee Powder Modules

The expanded role of the packaging design specialist is well demonstrated in this case history of the development of a bulk container system to ship coffee powder from extraction plants in Londrina, Brazil to the United States. The design criteria for the bulk containers in this example involved a great deal more than the containment and protection of the contents to be shipped. It included a comprehensive study of the entire logistical process, as well as finding ways to incorporate design innovations in the containers to meet the functional and economic objectives of the many activities as the containers moved from the supply source to the point of consumption.

The packaging specialist became involved in the design of the equipment to fill and discharge the contents from the containers, and in the adaptation of the container design to accommodate

handling and shipping without the use of pallets. He contributed
to the design of a vibrator system to deaerate the material during
filling to increase the product density, and thereby reduce the cost
of packaging per ton shipped. He built and tested prototypes,
assisted with shipping tests and cost evaluation of the bulk method, and followed the project through to implementation.

The system was developed and tested over an 18-month period of time by the Maxwell House Division of General Foods Corporation and implemented in 1982. The project was initiated by
a corporate staff man who specialized in the development of distribution methods. Prior to the coffee powder project, he had
achieved success with bulk container shipments of raw materials
such as bulk raisins, starch, and bread crumbs. He subsequently
made a presentation on the bulk container concept to a group of
Maxwell House division managers. A decision was then made to
study and evaluate the functional and cost feasibility of adapting
the bulk container method to imports of spray-dried coffee powder from Brazil.

There were many reasons why the Maxwell House people
wanted to improve the existing system. The powder was packed
into plastic polyethylene bags in corrugated shipping cases. Densities varied according to the type of powder, but typically averaged
82 net pounds per case. Unitized shipment of the cases on pallets
was economically impractical with this lightweight product. Consequently, all cases were hand stacked in and out of the ocean
containers, a process that took up to 12 man hours at both the
shipping port and the receiving locations in the United States. The
cases were relatively weak and did not stack well in warehouses.
Many collapsed, resulting in the bursting of the seams of the inner
liner bags and causing the leakage of powder. Opening and dumping the powder to receiving hoppers was a slow, tedious, and dusty
operation. Shrinkage was substantial, and the collection and baling of the liners and corrugated waste materials from many small
cases was time-consuming and costly (Figure 12.1).

Many people are involved in international shipping projects
of this kind. It was necessary to organize a task force consisting of
representatives of the various functional areas that would be in-

Corrugated shipping containers and plastic liner bags from local suppliers arrive at plant at Londrina, PR Brazil.

Crew of workers set up cases, glue bottom flaps, and install plastic liner bags.

Set-up cases are placed onto a conveyor belt and moved onto a scale under a filling head.

Case is filled with 82 net lbs (37.2 kg) of powder.

Cases move along conveyor to end of line, where they are loaded onto pallets.

Palletloads are moved to the shipping dock by lift truck and staged for loading.

Flat-bed truck arrives at dock and pallets placed on it by yard lift trucks. Load is secured and covered by tarp and dispatched to Port of Santos, Sao Paulo.

Palletloads of 40 kg cases are unloaded by lift truck and staged near container stuffing area.

A 40 ft ISO ocean container is trucked to the dock. Lift truck transfers palletloads one by one to the container doorway.

Crews of dock workers manually depalletize cases and fit them into the container. A total of 305 cases fill out the container. Takes 12–13 man hours time to complete loading.

The ocean container is closed and sealed and trucked to the dock.

Container is placed aboard ship by crane

Ocean trip to USA.

Container unloaded by crane, moved to customs inspection area.

Following release from customs, container is placed by crane onto a highway chassis and transported to receiving plant dock.

Container opened and the 305 cases loaded by hand onto pallets.

Pallet loads moved to storage area and stacked two high.

When scheduled, pallets taken to dump room and staged at dump station.

Cases opened, plastic bags' open ends folded over sides and banded.

Cases inverted by hand and contents shaken out into a receiving hopper.

Empty liner bags placed into pile for baling.

Empty cases stacked for compressing and baling.

Bales of bags and cases staged for pick-up by the salvage truck.

Figure 12.1 Traditional method of shipping coffee powder from Brazil.

volved. The committee included the engineering and operations managers of the plants, representatives of the coffee buying office and the broker in New York, the supplier in Brazil, and various staff services people from the company's packaging, quality assurance, transportation, and distribution areas.

Project leadership could have come from any one of the functional areas involved. In this case it was jointly shared by the Corporate Distribution Services area and an industrial engineering manager from the Maxwell House division.

The first step in the development process was to flow chart the existing system in detail. A tour of the entire system from the supply source in Brazil to the packing plant at Hoboken, N.J. was made. The system was discussed with operations, transportation, warehousing, and packaging managers, as well as brokers, buyers, and administrative personnel along the entire shipping route.

System flow charts need not go into great detail, but should provide enough information to help the specialists in all functional areas to understand what is involved. In this case it helped the packaging specialist to understand the performance requirements as the containers progressed through the system. For example, palletless handling by slipsheet was considered necessary all along the line from set up at the packing plant in Brazil to the discharge of powder at the receiving plant in the United States (Figure 12.2).

The palletless feature was essential to minimize the tare weight as well as to maximize inside cube utilization of the ocean container, since the product was relatively lightweight. The flow chart indicated that the tote boxes must be handled and moved a minimum of six times along the route. At one storage location they would be stacked four high. Preliminary specifications for the tensile strength of slipsheet tabs and the top-to-bottom compression resistance of the filled tote box were established to meet those conditions.

An incident occurred early in the planning process that is worth mentioning. Coincidental with the initiation of the project at General Foods, a large paperboard box manufacturer in the United States sent a packaging expert to Londrina to visit the supplier, Cacique de Cafe Soluvel, to discuss the feasibility of

EXTRACTOR PLANT IN LONDRINA

Dock container

Unload 400 totes in stacks of 16 each.

Stack of totes moved to process area by lift truck.

Manually set up a tote.

Install PE liner bag into the tote.

Set tote on vibrator table directly under fill spout.

Connect housing around tote.

Inflate airbags of table. Take tare weight reading. Connect end of bag to perimeter of filler spout.

Fill tote half full and vibrate 30 seconds. Fill full and vibrate 30 seconds. Overfill and vibrate until powder-level with top.

Deflate air bags to contact scale under table. Take out or put in powder to get precise 700 lbs (317.5 kg); take 3 oz sample. (85 gram)

Remove bag end from fill head and twist into tube. Twist the tube over and tape securely.

Install top cover.

Mark batch no., tare, net and gross weight on the tote.

Remove tote by push-pull attachment. Transport it to dock staging area.

Stack totes two high in staging area near docked container.

SAO PAULO, BRAZIL

40 ft ISO ocean loaded with 400 slipsheeted tote boxes is trucked 350 miles (563 km) to Londrina.

Sweep out the container Make sanitation check.

Lift truck with push-pull — move stack of two totes into container.

Load 21 stacks following configuration diagram.

Tape small box of samples on top last stack in doorway.

Close and seal container

Truck container to Port of Santos.

Ship to Port Elizabeth, N.J., USA.

Truck container to plant at Hoboken, N.J. and dock.

Lift truck with push-pull — unload and warehouse 4 high.

Move tote to process area and place on tilt table.

Remove cover, open bag Tilt tote to one corner Insert vacuum tube in tilted corner, pull vacuum to remove powder.

Collect and bale liner bags. Collapse and stack the empty totes.

Figure 12.2   Slipsheet coffee totes system.

using large, bulk paperboard containers for the export of coffee powder. Out of this came plans to fill four large pallet boxes with powder for shipment to the United States as a demonstration of the feasibility of the containerized bulk method. Although the proposed test was not part of the planned development program, General Foods agreed to receive and evaluate the test boxes.

The containers were RSC (regular slotted container) type, corrugated boxes, 44 in. (1118 mm) square and 36 in. (914 mm) high. They were mounted onto specially built wooden pallets that were 44 in. square and 6 in. (153 mm) high. Polyethylene liner bags were installed in each box that were then filled with 450–500 lbs (204–227 kg) of powder. The filled boxes were then secured to the pallets with steel straps, and shipped double stacked in the rear end of a 40 ft ISO ocean container as part of a regular shipment of 40 kg cases to the General Foods plant at Houston, Tx.

Upon arrival at Houston, the boxes were unloaded by a forklift truck and examined. The paperboard walls had been weakened and scuffed, but the contents were found to be in excellent condition. Although the demonstration shipment did not meet the development project objectives for General Foods, it helped to build credibility for the idea of shipping powder in large bulk containers. It further served to verify some of the problems anticipated in the original project plan such as:

> Pallets under the bulk boxes add substantially to packaging and shipping costs. The space they take up under the boxes reduces the total net weight of powder shippable in the same size ocean container in comparison with hand stacked 40 kg cases. The tare weight of pallets adds to shipping costs since ocean rates are based on gross load weights. A cost-effective bulk system would depend upon palletless handling, as anticipated in the development plan.

> The RSC-style design made very large boxes difficult to handle due to the collapsed dimensions. The 44 in. (1118 mm) square boxes flattened out to 88 in. (2235 mm) length and 62 in. (1575 mm) width.

Examination of the powder at Houston revealed still another problem. The boxes had been filled to the top when dispatched. Trapped air around the particles had seeped out in transit, leaving voids of several inches between the powder and the top of the box. A method would have to be found to deaerate the powder more completely during filling.

Flow charts will not provide the answers to all questions, but they can highlight the questionable areas for further study. For example, the flow chart (see Figure 12.1) indicates that shipment will be made in standard 40 ft ISO ocean containers. The boxes must, therefore, be designed to be modular to the inside dimensions of the 40 ft container. The inside length of such containers are 474 in. (12 m), the inside width is 92 in. (234 cm), and the height varies from 90 to 96 in. (229 to 244 cm). If the packaging design specifications provided for modules that precisely fit the inside dimensions of the ISO container, the filled totes could not have been fitted into the space available for several reasons:

The lift truck must transport a stack of two totes through the doorway of the ISO container. The overall stack height could not exceed the doorway clearance less a minimum of 4 in. (10 cm) for adequate operational clearance.

The totes would have to be sized to go through the doorways of the minimum 8 ft (2.4 m) containers, since the shipping lines could not guarantee the availability of the taller containers when needed.

The walls of the paperboard totes would contain some bulge when filled. It was not known at the time just how much bulge would result, since the filling methods had not yet been tested. This led to the decision to dimension the totes to be slightly rectangular, so that they could be loaded either in straight rows or in a pinwheel configuration (Figure 12.3) depending upon the amount of bulge.

KC TOTE	–Designated KC - the K for Kupersmit, the designer and C for coffee. Designed to transport lightweight material. Dimensioned modular to ISO ocean containers for export shipments. Collapsed dimensions are 47.5×42×5 in. (121×107×13 cm).
SLIPSHEET	–A high tensile solid fiber paperboard sheet coated with polyethylene on one side. Permanently bonded by glue to bottom of the tote. Two pull tabs, each 3 in. (8 cm) wide, one on short side and one on long side of the tote.
CAP	–The top cover of the tote. It is fastened onto the tote outer collar by plastic with a plastic clip on either side.
COLLAR/BASE	–The base is formed by flaps extending from the collar, to which the slipsheet is bonded. The walls of the collar fold upward from the base to provide a protective housing around the inner cell.
INNER CELL	–An independent, heavier corrugated section scored to collapse folded into the base. It provides the load-bearing part of the tote. 5 in. (127 mm) flanges extend around the top perimeter. To set up the tote, the outer collar is erected and the inner cell pulled up and unfolded to fill out the tote. Vertical scorelines are locked into place by a hingelike extension. Flanges are folded horizontally and taped down.
LINER BAG	–Gusset type, 4 mil, polyethylene bag dimensioned 46×42×110 in. (1168×1067×2794 mm) Noncling inner surface.
TARE WEIGHT	–Slipsheet tote with liner bag—65 lbs (29.5 kg).
CAPACITY	–Inside cube of set-up tote is approximately 1.2 cu m (42.38 cu ft).

Figure 12.3 Nomenclature—K box coffee powder tote.

Preliminary guidelines for an integrated bulk system as a result of the flow chart study included:

The tote box would be sized 47.5×42×43 in. (121×107×109 cm). That would allow a combination of adjacent lengths plus widths [89.5 in. (227 cm)] or two adjacent widths [84 in. (213 cm)]. The 43 in. (109 cm) height would allow a stack of two totes to be transported through the doorways of any ISO container.

The OD cube of totes with those dimensions would be 49.6 cu. ft (1.40 cu m). The inside cube would be approximately 46 cu ft (1.3 cu m). It was estimated that the packed density of the particular kind of powder would be about 12.4 lb/cu ft. That indicated a tote sized to provide 46 cu ft inside space would hold up to 570 lb (258.5 kg) of powder.

To stack the totes four high on warehouse floors, adequate top-to-bottom compression strength must be provided in all of the totes. Storage conditions, temperature, and humidity must be considered in design specifications. Complete information of this kind was not known at the time. Furthermore, there existed no previous standards nor guidelines on assurance factors for this kind of paperboard box. The decision was made to provide a four to one assurance factor for top-to-bottom compression resistance. A three to one factor would have been adequate under normal storage conditions, but with many unknowns at the time four to one appeared to be prudent. This simply means that the bottom tote in a warehouse stack must be capable of supporting up to four times the weight of the load carried on top. If the stack is four high and each tote has a gross weight of 600 lbs (272 kg), the design specification would be 7200 lbs (3266 kg) $[(3 \times 600) \times 4 = 7200]$.

To achieve cost-effective transportation, the net weight of powder in totes had to be at least equal to that in a shipment of 40 kg cases in the same size ocean container. A total of 305 cases filled a standard 40 ft ISO container. At 82 lbs net (37.2 kg) per case, the net weight per shipment totaled 25,010 lbs (11,344 kg). With the totes dimensioned to get a total of 40 per shipment, the required load per tote would amount to 625 lb each (284 kg). If bulge tolerances would permit, a pinwheel configuration would give 21 stacks of two, or 42 totes per shipment, which would lower the minimum to 595 lb (270 kg).

Space available in the filling area at the supply source, as well as the discharge area at the receiving plant, was limited. This, along with the set up and collapse time to handle the large tote boxes, was an important design consideration.

The type and sizes of corrugated board used in the construction of the boxes and the plastic liner bags would have to be obtainable in Brazil, since the cost of importing them from the United States could be cost prohibitive.

The success of the project depended mainly on the bulk container design. It must be adaptable to slipsheet handling and yet simple enough to set up and collapse. A decision was made to concentrate on the K box design. K boxes were designed and patented by Julius B. Kupersmit, President of Containair Systems Corporation of New York. These boxes have solid fiber slipsheets bonded permanently to their bases. Mr. Kupersmit agreed to build a prototype based upon the preliminary specifications established, and he agreed to put it through packaging laboratory tests to simulate warehousing and shipping conditions (Figure 12.4).

At that time, Packaging Research Laboratory at Rockaway, N.J., was one of the few certified packaging test labs that had equipment to test very large bulk boxes. The prototype was sent to the laboratory and put through a series of dynamic load tests as follows:

DYNAMIC COMPRESSION TEST NO. 1 (EMPTY TOTE)

Equipment used was a 30,000 lb (13,608 kg) Tinius Olsen Compression Machine equipped with an autographic recorder. The prototype was set up and placed empty into the machine. A preload of 250 lbs was applied, the autographic recorder engaged, and the test initiated. The upper platen was then lowered at a uniform rate of $1/2$ in. (1.27 cm) per minute until failure occurred. The results of this test found the empty tote was capable of sustaining a 10,000 lb (4536 kg) load at a 0.82 in. (2.08 cm) deflection before final failure. That indicated the design to be well within the four to one specification established for four-high stacking.

VIBRATION TEST

The equipment used was a 5-ton capacity, variable-speed vibration package tester with a circular motion of 1 in. (2.54 cm) double amplitude.

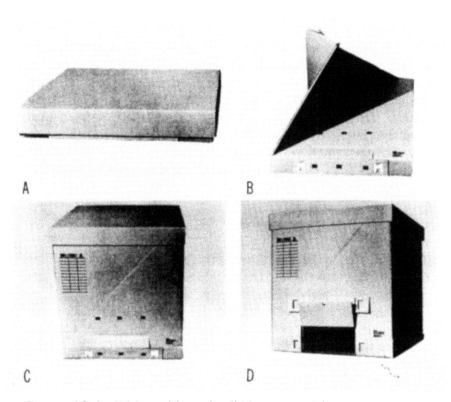

Figure 12.4 KC box with gravity discharge gate. Photo courtesy Containair Systems Corporation, Springfield Gardens, N.Y.

The tote was filled for this test. Since the actual product was not available, it was loaded with 800 lbs (363 kg) of plastic pellets. The filled tote was transferred by lift truck onto the vibrator deck, secured in place, and vibrated at 210 cpm for 1 hour. At this speed, the tote actually lifted off the vibration table momentarily at some interval during each cycle. Examination at the end of the test indicated that the container received only minor scuffing at the bottom. There was no serious damage nor leakage of contents.

INCLINE IMPACT TEST

Following the vibration test, the same tote was moved to the incline impact tester and placed on the machine's dolly flush with its leading edge. The dolly was then raised up a 10% grade to a distance of 43 in. (1092 mm) and released to roll down and crash against the back wall. Impacts from this distance resulted in a minimum impact velocity of 5.75 ft (1.75 m) per second. The tote was subjected to two impacts against each face. Examination of the tote following the test found no serious damage nor leakage.

DYNAMIC COMPRESSION
TEST NO. 2 (FILLED TOTE)

The same tote was then put through a final compression test. This time the filled tote sustained a maximum load of 9000 lbs (4082 kg) at 1.19 in. (3.02 cm) deflection.

Laboratory tests of this kind give an indication of how the tote may perform under actual conditions and may also help the designer to find potentially weak spots in order to make design modifications. Decision was made to produce four prototype totes, send them to Londrina, Brazil, fill them, and then put them through the entire shipping cycle for an ultimate performance test.

At that time, the supplier, Cacique de Soluvel, did not have a lift truck equipped with a push-pull attachment. The four filled totes were transported on pallets to the port of Santos, and there were manually pushed and pulled off the pallets and placed into position into the end of the ocean container. Like the previous test of bulk containers on pallets, the slipsheeted totes arrived in good condition. Again the powder had compacted, as trapped air seeped out in transit, and large voids over the powder were found when the totes were opened.

It was realized that the slipsheet method would require push-pull attachments to be installed on lift trucks at both the source and receiving plant locations. Unfortunately, the slipsheet method and push-pull attachments were not in general use in Brazil. Therefore, the procurement of an attachment and the training of a lift-truck operator to use it could further delay the project.

In any case, it was concluded that the slipsheet method would be critical to the project success. Consequently, provision for procuring the special push-pull attachments, installing them on the lift trucks at Londrina, as well as the training of the lift-truck operators in Brazil to use them, would be necessary.

The second and most challenging problem to overcome was the deaerating of the powder during filling. The initial test filling of the prototype indicated that the maximum net weight of the powder that could be filled into the cube available was an inverse function of the rate of input; that is, the greater the rate of input, the less the net weight, due to the amount of trapped air around the particles. The trapped air seeps out during transit, which leaves large voids over the powder by the time it arrives at the receiving location. The density could be increased by slowing down the fill rate to allow more time for natural deaeration, however, that would impact production operations unfavorably and was not economically feasible.

Samples of the spray-dried powder were shipped to a testing lab of the Cleveland Vibrator Company in Cleveland, Oh., to determine if the powder could be deaerated by vibration during the filling in order to increase density and thereby eliminate excessive compaction during shipment. Small samples were put through a series of vibration tests using a multiple frequency tester. The most efficient deaeration for this kind of powder was achieved with a vibration frequency of 1800 rpm.

The vibrator table was designed to have two electric rotary vibrators, one on each side of the table. They could be mounted either under the table or on the surface on either side of the tote box. The purpose of the two vibrators was to make it possible to control the direction of force into the box. The vibratory force is created by eccentric weights rotating off the motor shafts. By rotating the weights on one vibrator motor clockwise and rotating the other counterclockwise, it is possible to cancel out horizontal forces and to concentrate a vertical vibratory force up into the tote box. (Figure 12.5).

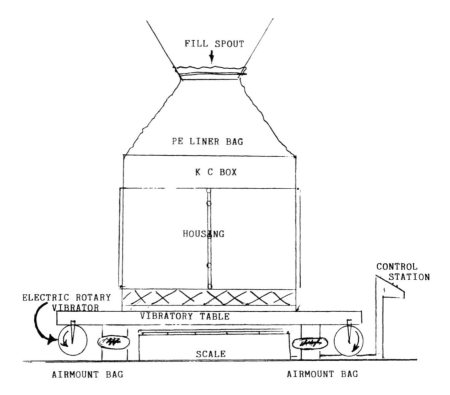

FILL SPOUT

PE LINER BAG

K C BOX

HOUSING

CONTROL
STATION

ELECTRIC ROTARY
VIBRATOR

VIBRATORY TABLE

SCALE

AIRMOUNT BAG   AIRMOUNT BAG

Figure 12.5 Vibration densifying system. PE liner bag is installed into a KC box and box is set onto the vibratory table. Open end of liner bag is banded onto the powder discharge spout. A metal or wooden housing is clamped around the box to keep its walls rigid during vibration. Electric rotary vibrators are suspended from either side of the table. They could be mounted underneath, as shown above, or on the topside. Eccentric weights on one are rotated counterclockwise and on the other one clockwise. Airmount bags are inflated to isolate the vibration assembly from the floor. When deflated the load comes into contact with scales for weight checks. Powder can be added or scooped out to get a precise net weight.

Final design specifications for the vibrator system based on the frequency selected were then developed as follows:

Load weight: table deck — 425 lb (193 kg)
            two vibrators — 186 lb (84 kg)
            tote tare wt — 60 lb (27.2 kg)
            powder weight — 700 lb (317.5 kg)

Twin rotary electric motors with 1800 rpm ratings

Output force (from motor manual)=2640 lb (1197.5 kg) each vibrator for total of 5280 lb (2395 kg)

Constant (from manual)=70,470.91

Amplitude=total output force×70,470.91÷total load weight×frequency squared.

[Loaded amplitude calculates to 0.084 in. (0.213 cm) and unloaded amplitude to 0.188 in. (0.47 cm).]

G's=output force divided by total load weight. (calculates to 3.85 loaded and 8.64 unloaded)

The deck of the vibrator table is best suspended on air-bag isolation mountings. Not only does this result in less energy loss during vibration, but functionally it can be used in conjunction with a scale. The air bags are inflated fully to raise the deck above the scale during vibration and then are deflated to rest the entire load on the scale for weighing. With this vibration method, deaeration can be very fast, requiring only intermittent bursts of 20 to 30 seconds during the filling process. Experimentation with the filling of coffee powder at Londrina indicated the best procedure to be:

1. With air bags deflated and the tote resting on scales, record tare weight and reset scale to zero.
2. Fill the tote about half full.
3. Inflate the air bags and vibrate 20 seconds.
4. Fill to the top.
5. Vibrate 30 seconds.

6. Fill until the powder forms a mound well above the top of the tote.
7. Vibrate until the powder levels off within the top of the tote.
8. Deflate the air bags so that the tote is in contact with the scale. Scoop out or add a little powder to get the precise net shipping weight wanted for all totes.

The vibration method increased the density of the packed powder approximately 35%, which meant 35% fewer totes and container shipments required for the same amount of powder exported. Slowing the filling rate would have allowed more time for natural deaeration during loading, but that was impractical. The following comparative densities chart illustrates the effectiveness of the vibration method compared with normal and reduced filling speeds.

Fill an ID Cube of 42 CU FT (1.12 CU M)

Normal fill rate	12.4 lb/cu ft	520 lb/tote
		236 kg/tote
Fill at half speed	14.2 lb/cu ft	596 lb/tote
		270 kg/tote
Normal fill plus vibration	16.7 lb/cu ft	700 lb/tote
		318 kg/tote

As the first full containerload of 42 totes were to be shipped from the supplier plant at Londrina, the need for the packaging designer to stay with the project throughout the initial filling and shipping operations became apparent as the project reached the trial shipment stage.

Unforseen problems inevitably occur in projects of this kind. The most critical during the filling process was the bulging of the paperboard tote walls. Operation of the vibrator compacted the powder rapidly, causing pressure to build against the walls. This resulted in the bulging of up to 2 in. in both the length and width dimensions, which was serious enough to limit the number of totes that could be fitted into the standard ISO ocean container.

Mr. Kupersmit, the tote designer, had two options. One would be to strengthen the paperboard walls to resist excessive bulging. That, however, would have added to packaging costs and tare weight of the tote. The second option was to provide a means of containing the amount of bulging during filling by mechanical means. At his direction, a rigid plywood girdle was built to fit tightly around the tote during the operation of the vibrator system. The rigid girdle kept the paperboard walls of the tote from deforming as compaction took place. When the tote was filled to capacity, lateral forces inside were equal against all walls, and the girdle could be removed as the walls remained in place without bulging. The method worked well, and at a later time the plywood girdle was replaced with a cabinetlike enclosure made of metal.

A few other less serious packaging and handling problems surfaced during the preparation, filling, and loading of 42 totes for the initial containerload shipment. Most of these were solved on the spot. As examples:

> The cells of the K box tote nest inside the base of the collapsed, empty tote box. To set up K boxes, the top cap is removed from the collapsed box and the outer collar pulled up into place. The usual practice with shorter K boxes is to reach down inside the collar, grasp the collapsed cell, and pull it up to fit inside the collar. The 43 in. (1092 mm) tall coffee tote box, however, put the collapsed cell in the base out of easy reach. Workers leaning over the sides, and stretching to reach it, caused damage to the first tote set up. A long handle with a hook on its end was quickly fabricated in the plant shop and solved the problem.

> When the tote is set up, a polyethylene liner bag must be installed in it. The bag specifications, which had been established in earlier tests were to be 4 mil polyethylene, gusset type, 46 in. (117 cm) long, 42 in. (107 cm) wide to fit the inside dimensions of the tote, and 110 in. (280 cm) tall in order to have adequate length to fold and tie over the filled load. The simple stuffing of the end of the bag down into the tote would have left many wrinkles and folds that could form air pockets and reduce the load capacity. The

immediate solution was to affix a wire plate onto a long handle. The plate was sized slightly under the inside dimensions of the tote, so that it could be inserted into the bag and pushed down inside the tote to position the bag in place. Later someone got the idea of manually forming the open end of the bag into a tube and inserting a wand from a blower to inflate the bag inside the tote with filtered air.

To fill the first tote, the long end of the liner bag was folded down over the outside of the box and held in place by an elastic rope to prevent it from being pulled down into the box during filling. The open top, however, allowed considerable dust to escape during the filling process, since the powder fell 4–6 ft (1.2–1.8 m) from the fill spout into the tote. The dust not only created an unpleasant work environment in the filling area, but resulted in a relatively high shrinkage loss.

A modified filler head was subsequently fabricated in the shop and installed. It provided a wide circular shield with a metal belt fastener to hold the open end of the liner bag. A pressure relief valve in the shield provided an escape for displaced air as the filling took place.

Push-pull attachments to handle slipsheeted loads were not in general use in Brazil at the time, and of the few to be found in Brazil, none were in operation anywhere near the Londrina area. It would have taken weeks or months to have a new push-pull attachment manufactured in the United States and shipped to Londrina. Through the cooperation of the representative of Cascade Corporation in Sao Paulo, an old, used push-pull attachment was located and put in working order. It was brought to Londrina and installed on a pneumatic-tired, yard lift truck, the only lift truck available that had a low enough mast to allow entry through the doorway of an ISO container. It is much more difficult to operate a lift truck with a push-pull attachment compared with handling pallets with fork tines. Training films were sent to Londrina for viewing by the lift-truck operator selected to load the initial shipment. He was given time to practice the gripping, pulling, stacking, and discharge functions to build up sufficient skill to handle the loading operation. Despite the newness of the method, the operator was able to do a creditable job of loading the ocean container after just a few hours of practice operating the push-pull attachment.

It was essential that the ISO intermodal container be brought from the Port of Santos at Sao Paulo to the supplier's plant at Londrina, a distance of some 350 miles (563 km). The 40 kg cases were shipped by truck to the port and loaded there into the containers by hand. The totes were too big and cumbersome for manual loading, and push-pull equipment did not exist at the port. Skeptics back in New York were convinced that it would not be possible to transport a large 40 ft ocean container inland in Brazil. They had assumed highways in that part of the world would be primitive with limited overhead clearances and inadequate roadbeds for the weight of the load to be transported. Actually, an excellent highway existed, and the ocean container trucked over it without difficulty.

Docking the container to permit a lift truck to enter at deck level required a ramp to be built over the existing low dock at the plant. A simple ramp with a 10% grade was built of wooden planking in the plant shop and worked well.

The load configuration necessary to get 42 totes into the container was plotted (Figure 12.6) and then furnished to the shipping dock foreman. Loading was carefully monitored to ensure that the stacks were properly positioned and compressed tightly together. The totes were dimensioned to fit snugly and to fill out the entire container, which eliminated any need for dunnage to stabilize the load.

Intermodal containers are closed and sealed at the point of dispatch and not opened again until they arrive at the port of destination, which in the case of this initial shipment was Port Elizabeth, N.J. The USDA procedure for port inspection of containers of hand-loaded 40 kg cases is to open the container and take out a few cases for inspection prior to releasing the shipment from customs. A container arriving with large heavy totes on slipsheets could not be removed and inspected, since the port had

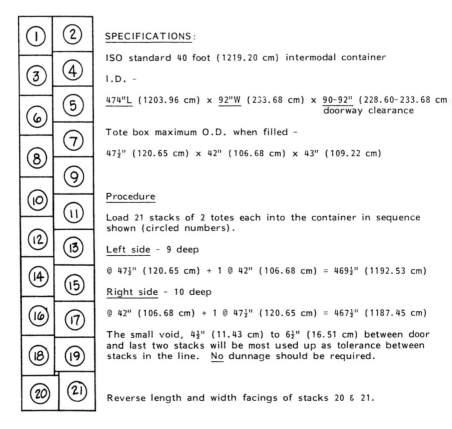

Figure 12.6   Loading diagram for spray-dry coffee totes.

no lift trucks equipped with push-pull equipment at that time. In anticipation of this problem, small samples of powder from each tote were taken after filling and collected in a small box. The full box of samples was taped to the top of the load just inside the doorway. The USDA agents at Port Elizabeth were, therefore, able to inspect the doorway totes and box of samples, which was adequate for the release of the container to its final destination.

Upon arrival at the Hoboken, N.J., plant of General Foods, the container was backed against the receiving dock, opened, and a lift truck equipped with a push-pull attachment removed the doorway stacks. These first totes were opened and samples were taken for evaluation in the plant quality assurance lab. The load was acceptable to the quality assurance manager, and the shipment was released for unloading and warehousing.

The productivity gain of mechanical unloading versus hand unloading 40 kg cases was substantial:

Comparative Productivities – Unload 40FT ISO Container

	40 kg cases	Slipsheet totes
No. per container	305	42
Unload and warehouse (man hours)	12	1
Net weight of powder		
lb	25,010	29,400
net kg	11,344	13,336

The final step in the logistical process was to remove the powder. For the old method, the 40 kg cases were opened, and the powder was hand dumped into receiving hoppers. Mechanical methods had to be developed to remove it from large totes that contained 700 lb (317.5 kg). The simplest method would be to invert the tote mechanically and to let the powder flow out the top. Control of the rate of flow and dust was the main concern with such a method. A decision was made to remove the powder by vacuum.

A metal table, designed to tilt to one corner, was installed in the delivery room. A large tote was moved into the room by a lift truck with a push-pull attachment and placed on the tilt table. The tote cap was removed and the liner bag opened and folded down over the sides. A stainless-steel wand connected to a suction hose was pushed down into the powder in the corner to be tilted. The entire tote is then tilted on the corner to approximately 50°. Suction can be created by either introducing the powder into an air stream or by pulling a vacuum in a chamber that intermittently fills and releases powder through a rotary airlock (Figure 12.7). It is a very clean operation, with little dust escaping during the process. Shrinkage in the discharge area, while high with the hand-dumped case system, was virtually eliminated with the bulk totes.

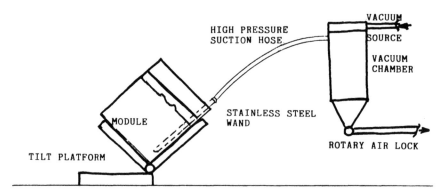

Figure 12.7   Vacuum discharge method.

(1) Bulk module is placed on tilt stand. (2) Top removed from module, inner liner bag opened and folded down along sides. Elastic band used to keep liner in place during discharge. (3) Module is mechanically tilted to one corner. (4) Stainless-steel wand connected to a high-pressure suction hose is inserted into the material in the tilted corner. (5) Vacuum is pulled into the vacuum chamber. (6) Material flows in air stream into the chamber. (7) Airlock valve at the bottom of the chamber releases the material to process.

When all of the powder was removed, the liner bag was taken out and placed into a collection box for later baling and disposal. The tote box was collapsed and stacked for sale as surplus, or reused for the transport of blended powder to other plant packing locations.

No matter how successful the physical aspects of such a project are, the new method must stand up costwise in comparison with the old method. Many studies were made to determine the bottom line cost comparison of the old 40 kg case method versus the bulk tote system. In the process it was found that production of the slipsheet totes in the United States for shipment to Brazil was exceptionally costly. It would be cheaper to produce the totes and slipsheets in Brazil if a suitable supplier could be found. The tote developer, J. Kupersmit, subsequently visited paperboard suppliers in Brazil, and he found a supplier that had the necessary equipment to produce large corrugated boxes and slipsheets. A licensing agreement was worked out between Containair Systems Corporation in New York and Rigesa Ltd., Sao Paulo, Brazil to produce the coffee totes. A source of supply of 4 mil polyethylene liner bags was also found in Sao Paulo.

The tote box project was implemented in 1982 and quickly spread throughout the industry. Within a couple of years the volume exported from Brazil had grown to over 15,000 tons annually, and continued to grow as other suppliers and other packers joined the program.

With the bulge control method perfected, it would have been possible to increase the width dimension of each tote up to 2½ in. (63 mm), which would have increased the cubic capacity enough to get the same net tonnage each shipment into 40 totes instead of 42. Additional savings would have been possible with fewer totes to handle through the process, and one container shipment could have been saved for every 20 shipped. The savings would have certainly exceeded the incremental packaging costs of the slightly larger totes. Nevertheless, the original tote size had been well integrated into the total system, and the project was economically sound. No one felt the additional savings that could be realized by changing the tote dimensions was really worth the time and effort required to implement the change, and the original dimensions became standard.

EXPANSION OF BULK TOTE METHODS TO COLOMBIA

In 1987 the feasibility of importing agglomerated coffee powder in the bulk slipsheeted totes was being studied at General Foods. The supplier, Col Cafe of Medellin, Colombia, S.A., had exported the powder in corrugated cases holding up to 45 kg (99.2 lb), but large bulk totes had never been tested for agglomerated powder.

The packing and shipping knowledge gained on the Brazilian project saved a great deal of experimentation with bulk methods in Colombia, but the fragility of agglomerated powder made it necessary to make design modifications to the bulk tote. Agglomerated powder is made up of larger particles or clusters of small particles.

These particles can disintegrate into dust fines if subjected to severe handling impacts. Densification by vibration during filling would have to be carefully controlled. The discharge of the powder on the receiving end would have to be as gentle as possible. The vacuum removal method could not be used, since the large particles could break down in a high-velocity air stream.

The initial work on the project did not involve packaging specialists. It required the work of food science experts to develop manufacturing specifications to get the right combination of particle size, color, firmness, and hardness characteristics that would be acceptable for marketing, as well as adaptable to packing and shipping the powder in bulk packs.

High-density packing of the product into the totes was a matter of controlling the vibratory forces during filling. The twin rotary vibrator system developed for the densification of spray-dried powder was easily adapted to the agglomerated powder. It provided deaeration without particle damage during filling through the adjustment of the force output and the duration of the vibration cycle time.

In order to remove the powder from the totes, modifications to the packaging design became necessary. With the smaller cases of this product, the practice had been to hand dump the powder from the top of the case, as gently as possible, into a large portable metal bin for intermediate storage. To discharge the powder, the metal bin was moved onto a mechanical tilt station in the process area and tilted slightly. A gate located at the base of one side of the bin was opened, and the powder was allowed to flow by gravity into a receiving hopper. When empty the metal bins were steam cleaned and used over and over.

The feasibility of building a similar side gate into the base of the paperboard tote was studied. There would be an added benefit, since with a side gate in the transit tote, the need to transfer powder into a metal bin for intermediate storage would be unnecessary. The investment in metal bins, and their maintenance and cleaning costs, would be eliminated.

To design a side gate into the K box tote, it was necessary to cut openings in both the outer collar and inner cell. The challenge for the designer was to find a way to provide openings large enough for the effective discharge of powder, which could be sealed securely to prevent bulging and leakage during transit.

An outer gate flap was cut into the tote collar and secured into place by means of two plastic clip fasteners. To open, the clips are removed and the flap pulled up and held against the tote wall by two small patches of Velcro. A similar but slightly smaller gate flap is cut into the inner cell. During transit, this flap is secured closed by the pressure between the product inside and the outer collar on the outside. A small Velcro strip is used to hold it open against the collar flap (Figure 12.4).

To discharge powder, the tote is positioned on the same tilt device used for metal bins. The inner and outer flap gates are opened, and the tote is tilted in the same manner as a metal bin. A knife is then used to cut the inside plastic liner bag, and the powder flows out by gravity to the receiving hopper. The side-gate gravity-discharge method may not be suitable for powders that compact tightly together, but the larger particles of the agglomerated product had sufficient lubricity to flow easily in a steady stream. It is possible to pack material that does not flow easily into a side-gate tote box and to use a vibrator to break up the mass inside for better flowability during discharge.

In that case, the added cost of providing a side-gate discharge in the tote, in comparison with some other mechanical means of removing the material, must be considered. With sufficient ongoing volume, the avoidance of accumulative incremental costs of side gates may easily pay off an airveyor system or mechanical inverter for materials that are not sensitive to damage during discharge by those methods.

The project to export agglomerated powder from Colombia provides an excellent example of the role of packaging in a total logistical system. Many independent areas contributed to the success of the project. The food scientists produced a product that was capable of bulk handling, and the transportation area worked on the availability of transport equipment and rates. Plant engineering people at the supplier's plant installed and adjusted vibrator equipment to achieve a dense pack. The shipping dock and warehousing personnel provided equipment and the training of lift-truck operators to handle and store slipsheeted totes. Operations personnel at the user plant developed the system for warehousing the powder in the transit tote and moving it directly to process. Throughout the entire chain of activities, packaging design was critical. To begin with, it had to be sized properly in order to be moved through filling, storage, and discharge areas and then fitted into tight-load configurations in the ocean containers. It had to be simple to set up and collapse when empty to minimize labor time and costs. Like the Brazilian project, the built-in slipsheet was mandatory for cost-effective shipping. Unlike the Brazilian project, provision had to be made for a built-in method of discharging the powder. Finally, it had to be a durable container that could stand the forces of transit and still be capable of being stacked up to four high for good utilization of warehouse space. Above all, it had to be affordable and had to minimize the packaging, handling, and transportation costs per ton shipped compared with conventional smaller cases.

This case history leaves little question on the need for packaging designers to become involved in the planning and implementation of the total logistical system for projects of this kind.

13

Case History 3: Top-Lift Modules

This case history describes the development and evolution of a bulk packaging system. Although the concept first appeared in the 1950s it is not widely known, and shippers who are unaware of its existence do not ask specifically for it. The intent of this case history is to thoroughly familiarize bulk packaging designers with the method and its potential, so they may be in a position to recommend it as an alternative system for certain kinds of products that are packaged and shipped in bulk containers.

Top-lift bulk containers have been designed in different ways depending upon the particular product. In all cases, the container design is interdependently related to the materials-handling equipment and handling methods. The packaging specialist must be thoroughly familiar with the materials-handling equipment and the top-handling technique in order to design a top-lift bulk container.

The top-lift concept can be demonstrated by opening a typical regular slotted container (RSC)-style corrugated case of goods and folding the flaps completely over and down the sides of the case. While holding the flaps in that position, lift the case straight upward with your hands. With large bulk containers that are too heavy to lift by hand, the flaps are flanges extending from a top cover that interlock with flanges that extend from the walls of the box. Lifting is done mechanically by steel blades installed on a lift truck or handcart (Figure 13.1).

A brief explanation of how the method originated in the 1950s and the early applications will provide the packaging specialist with an understanding of how and why this concept has been used over many years, as well as the kinds of products that are adaptable to it.

The idea for top-lift, or top-cap, handling, as it was known in the beginning, was developed at General Electric's Appliance Park in Louisville, Ky., in the early 1950s. According to Ralph McGaughey, who was General Electric's Facilities Planning and Project Engineer at the time, the concept evolved from handling

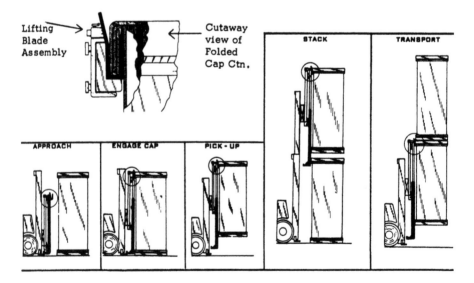

Figure 13.1   Courtesy of Basiloid Corporation, Elnora, Ind.

and packaging methods used to handle refrigerators at the company's plant at Erie, Penn., prior to the construction of the Louisville complex.

He recalled that the refrigerators were packaged in large corrugated cartons that were designed cap-and-tube style with top caps interlocked with the main tube, much like the top-lift cartons of later years. Horizontal handling was done by conventional two-wheel handcarts, and cartons were stacked two or more high in the warehouse by means of an overhead crane. Workers would hand truck the cartons to a crane pickup area and line them up. The crane operator would then lower a cable to which a hook was attached and insert the hook under the top fold of the caps. The cartons could then be lifted and stacked on the warehouse floor.

When the manufacture of refrigerators was moved from Erie to Appliance Park in 1953, other appliances such as ranges and home laundry washers and dryers were already being produced there. Those products were packaged predominantly in wirebound crates. Lift-truck tines were inserted through the tops of the crates for lifting and transport.

The consolidation of manufacturing operations for a number of different appliances at Appliance Park led to a management decision to standardize the packaging and handling methods of all appliances. The top-cap corrugated carton appeared to be the most promising method.

In order to integrate the top-cap method into the system, it was necessary to adapt it to handling by mobile lift trucks. General Electric engineers initiated a search for a metal fabricating company that would cooperate on the development of the lift-truck top-lift attachments. This led them to the Basiloid Corporation, a small company located in Elnora, Ind., which was formed during the war years to produce turrets for Army tanks.

In 1955 an agreement was reached with Basiloid's President, Lloyd Hobson, which resulted in a cooperative development program that lasted for 7 years.

Mr. Hobson designed a simple slotted frame attachment for lift trucks to which specially designed lifting blades could be easily inserted. The frame also functioned as a back plate to keep the carton upright during lifting. Subsequent refinements included a slide extension that would telescope out of the main frame to brace very tall cartons, forks that folded up into the back plate for fast conversion from top-lift to pallet handling, and frame width extenders to handle very wide cartons or wide loads of two or more cartons.

With the new equipment and improvements in carton design to strengthen the top folds, the system was installed in General Electric's appliance distribution centers throughout the country in the 1960s.

In the years that followed, Basiloid Corporation exhibited the method at materials-handling trade shows. With a Basiloid attachment bolted to a lift truck, Lloyd Hobson would hoist a large appliance carton high up in the air. Sometimes as a demonstration of the safety of the method, his wife, Lee, would climb atop the carton and ride up with it. Suspended high up, she would take a pair of scissors and cut the tape that held down the top folds. The simple demonstration showed that the function of the tape was to hold the resilient top fold down so that the lift blade could be inserted under it. Once the blade was in place, the top fold could not spring up, and the restraining tape was no longer needed.

The Basiloid trade show demonstrations, and a few trade press articles from time to time, were the only sources of information about the top-lift method that was available to packaging specialists. As others heard about it, the method was expanded to the packaging and handling of different kinds of products, including large pieces of furniture, bulk cartons of raw materials, textiles, and many other products. Basiloid took orders for lift blade equipment for shipment to a number of foreign countries. Such orders came through U.S. brokers, and exactly where the equipment wound up, and exactly what kinds of products it handled, was in most cases unknown to Basiloid. The simplicity of the equipment made unnecessary the need for training, follow-up, and parts service on the part of the manufacturer.

One unique application in the United States involved the handling of 640 lb (290 kg) blocks of cheese. The top-lift packaging was designed by the Weyerhaeuser Company of Manitowoc, Wis., as an alternative to wooden crates. The procedure, as outlined in Weyerhauser sales literature, is as follows:

A solid block of cheese, or a unitload of several smaller blocks of cheese, is fitted with a polyethylene bag and positioned on a corrugated tray that has extended flanges.

A laminated corrugated liner wrap is then placed around the block and taped into place.

The main carton tube, which has extended flanges top and bottom, is then fitted over the liner wrap.

The flanges of the tube are interlocked with the flanges of the bottom tray and secured with a perimeter tape.

The poly bag is sealed at the top and the top cover is placed over the block, and its flanges are interlocked with the flanges extending from the tube. The perimeter tape is then used to lock the top fold into place.

The 640 lb blocks of cheese were handled two at a time by Basiloid attachments on lift trucks and stacked five high in storage areas. In comparison with the use of wooden crates on wooden pallets, the total system costs were reduced over 65%. Key economic factors included reduced packaging, assembly time, and transportation costs. The tare weight per carton compared with wood was reduced from 111 lbs (50.3 kg) to 21 lbs (9.5 kg).

As indicated in the above example, the main reason for interest in the top-lift method has been the functional and economic advantages it has in comparison with other packaging and handling systems. The advantages include:

The attachments have no hydraulic connections, nor complex mechanical parts. Consequently, they cost but a fraction of other kinds of palletless handling devices, such as clamps or push-pull attachments, and the maintenance is minimal.

The attachments are relatively lightweight, and thereby adaptable to lift trucks with little loss of carrying capacity.

The turning radius of the vehicle is substantially less than that of lift trucks carrying forks or clamps that protrude out front. This, along with elimination of the need to leave access gaps around the stacks, provides exceptional high-density storage capability for the economical use of space.

The use of the top-lift devices is so simple that very little training is required of lift-truck operators. Learning curves require only a few minutes compared with days or weeks for other kinds of equipment.

In the days that preceded unitized shipping on pallets or slipsheets, the top-lift method made the mechanical loading of large cartons into transport vehicles possible. Today it can be used in conjunction with pallets or slipsheets.

Answers to questions on how heavy a load can be lifted by the top-lift method are usually more related to the economics of providing adequate tensile strength in the top fold, rather than the heavy lift feasibility. Most applications involve loads weighing no more than a half ton. For raw materials or other products in which the size of the carton is dictated by the dimensions of the product, the cost of strengthening the carton and its top folds for top-lift handling must be measured against the cost of reducing the size of the carton to smaller modules. The smaller bulk containers still offer an alternative to handling many conventional-sized cases of materials, and two or more can be handled at one time in clusters by installing multiple lift blades on the slotted frame attachment.

In 1985 a unique top-lift carton experiment was undertaken at General Foods that furthered the development of top-lift methods. New and larger sales outlets for the company's products were appearing that were known in the trade as mega stores, warehouse stores, and wholesale club stores.

Supermarkets, too, were growing in size, and the volume of goods sold was increasing. The rapid turnover of stock that characterized these kinds of giant markets required the purchase of many products in palletload quantities. Large crews of laborers were kept busy at the stores continually opening and cutting down thousands of small cases to build sales floor displays, or to stock shelves with retail packages.

The relatively small-sized cases of packaged goods facilitated the manual handling and stocking of shelves at the small stores, since they usually ordered the goods in one- or two-case quantities. That had become a burdensome and costly nuisance and an obstacle to improved productivity at the high-volume sales outlets.

As a customer sales service to the larger sales outlets, several large manufacturers, including General Foods, offered high sales volume products in prebuilt floor displays. These displays were usually built at the manufacturer's regional distribution centers.

There, cases of the products were opened and packed by hand into unitized displays on small wooden pallets. The displays were stabilized on the pallets by upright corrugated corner posts and overwrapped with stretch film. Since displays of these kinds travelled relatively short distances from the regional warehouses to the store docks, the rather flimsy protective packaging was adequate. On a total system economic basis, however, the packaging represented a substantial additional cost per package displayed, and the labor to open the original small cases and build the displays was simply transferred from the store to the regional warehouse.

It was recognized that total system productivity and costs could be vastly improved if the retail packages could be automatically packed into trays and assembled into displays on plant production lines. The displays would then be shipped in truckload quantities to regional distribution centers and distributed to retail stores for setting out on the retail floors, or for the convenient transfer of trays of packages to shelves.

The concept seemed worthwhile testing, and subsequently a product line of five different kinds of breakfast cereals was selected for an initial test in cooperation with a large warehouse store outlet. Since the products were normally purchased in palletload quantities of 20 to 30 small cases, the plan was to pack the equivalent number of packages in one large box. The huge box would have a top cover and front panel that would be removed for the display of the merchandise on the retail floor.

Although the products were relatively lightweight, the large containers would weigh several hundred pounds each, which dictated the need for mechanical handling. Mechanical handling was not a problem at the production plants, nor at the regional distribution centers. The use of slipsheets was to be used for transport from the plants to the regional warehouses. The slipsheet loads would then be transferred onto pallets for the short-haul deliveries to the stores. Most large stores had fork-lift trucks or pallet jacks on their receiving docks to remove the boxes on pallets. The display boxes were to be transported on pallets to the retail floor location by pallet jacks, then placed into position at the display location, and then the top cover and front panel removed.

During visits to several store locations to observe existing methods, it was noted that pallets, while essential to the storage of goods in racks, were a nuisance. Thousands of empty pallets had to be stored in yards, making them vulnerable to contamination and unsuitable for use under food products. The quality of pallets received with loads at the docks was generally poor, and most of the goods received on pallets had to be repalletized on captive house pallets for storage in the racked area of the warehouse. The maintenance of house pallets, and the replacement of broken boards, was a continuous added expense.

As a result, a decision was made to attempt to develop a totally palletless system for the plant-built display boxes. The extension of the use of slipsheets for the transport of the big boxes from the regional warehouse to the store appeared, however, to be impractical. The investment in push-pull equipment and larger lift trucks to carry them would put an economic burden on the stores. Furthermore, training lift-truck operators to use push-pull equipment would be a problem. Most of the workers on the store docks were transient employees, students, and others who would not be around long enough to develop the skills necessary for efficient operation of the sophisticated push-pull equipment.

The top-lift method seemed to make sense for this kind of system. Top-lift cartons could be handled palletless on slipsheets for the long haul from plants to regional centers and then handled by top-lift equipment at the store docks, as well as for positioning them on the sales floors.

The cereal products selected for the initial test appeared to be ideally suited to top-lift bulk containers, since they were relatively lightweight. A decision was made to proceed with the building of prototype boxes for the five cereal products. The Boise Cascade Corrugated Container plant at Marion, Oh., agreed to participate in the development program. Palletloads of each of the cereal products were shipped to the plant at Marion, Oh., Basiloid Products cooperated by sending a lift blade attachment to the plant for preliminary lift tests.

The large boxes ranged in size from 39×34×39 in. to 49.5× 41.5×48 in. (99×86×99 cm to 126×105×122 cm). Each carried the number of retail packages shipped in 16–32 conventional cases of the products. Gross weights per box ranged from 520 to 760 lbs (236 to 345 kg).

A bottom that will not collapse under pressure of the inside load is essential to top-lift cartons. Boise Cascade elected to use a slipsheet base with wings extending from the center, a method used with pallets in Europe to secure large containers onto pallets without the use of straps or nails. The wings were fitted through a gap between the bottom flaps of the box. They were then folded down on the inside to lock the bottom flaps into place. Boise Cascade called the method Pallet Pak.

The bulk containers were designed with top folds on all sides, and a precut front drop panel, which facilitated both packing and retrieving retail packages.

Packages were hand loaded into the big boxes for top-lift tests. With a Basiloid attachment installed on a lift truck, the boxes were lifted from all sides, transported, and stacked. They were lifted in stacks of two by inserting the lift blade in the top fold of the bottom box and allowing the top box to ride piggyback.

The five prototypes were then shipped to a regional distribution center, and from there to a warehouse store in Tulsa, Okl. They travelled through the system from plant to store display location without incident. Basiloid brought a blade attachment to fit onto a store lift truck and demonstrated its simplicity by installing the attachment in a matter of a few minutes. A lift-truck operator had little difficulty handling the boxes with it after only a couple minutes of instruction.

This particular store normally stored pallets of goods in reserve stock on pallets on rack beams over the display floor level. It was demonstrated that captive pallets could be left on the racks, or boards could be permanently mounted across the rack beams to provide shelves on which the top-lift boxes could be placed. That would avoid having to transfer the large boxes onto pallets at the docks for storage.

Despite the success and simplicity of the packaging and materials handling system, the size of the boxes, approximately 48 × 40 × 48 in. (1219 × 1016 × 1219 mm) tall, had certain retail sales functional problems.

To begin with, the wide displays limited the number of different products that could be exhibited in a line on a sales floor. Small cases could be stacked in columns on the pallets to exhibit several different products in a single pallet facing on the sales aisle. The large boxes containing a single product each committed the entire pallet space to that product.

Secondly, shoppers found it inconvenient to reach into large boxes up to a meter in depth to retrieve packages. Packages in the bottom tiers were especially difficult to reach. It was concluded that the large display containers should be sized no more than a quarter of a standard pallet load, i.e., quadrant modules. Quadrants would display all packages within easy reach [21 in. (53 cm) maximum depth]. Two quadrants with different products in each could fit into the same aisle facing as one pallet-size box. The height of the quadrant boxes would be governed by the density of the product and the need to position most of the packages high enough for shopper convenience. A box height limit of 48 in. (1219 mm) was necessary so that lightweight products could be double stacked in standard high cube trailers to minimize transportation cost penalties. Ideally, the optimum for a display box would be 24 × 20 × 45 in. (610 × 508 × 1143 mm) and 250 lb (113 kg). Such boxes, in unitloads of four each, stacked two high and two across in standard highway trailers, could be loaded 10 stacks deep in a standard 42 ft (12.8 m) highway trailer to achieve an economical 20 ton payload. The large variety of goods and densities of the products, of course, made such an ideal shipping condition impractical.

Palletless shipping of the quadrant modules was also critical to the system. Standard 48×40 in. pallets would reduce highway payloads by their tare weight, and the space taken up under the boxes would necessitate reducing the box height in order to stack two high in the transport vehicle. The use of special quarter-size pallets under each module would not only add to packaging costs, but would create a monstrous problem in handling and storing the huge numbers of small pallets should the quadrant display come into widespread use.

Slipsheet palletless methods could be used to transport unitized boxes in groups of four through the long-haul shipments from plants to regional distribution centers, and the slipsheet loads could be transferred onto pallets and shipped just one high in trailers to nearby stores. The problem would be handling quadrants that could weigh up to 250 lb (113 kg). They would be too heavy for manual handling and inconvenient to transport with a conventional two-wheel hand cart.

While the economy of scale benefits would be diminished with top-lift quadrants, it was decided to run a second test with the products packed in top-lift quadrant boxes. Again, the products and equipment were sent to the Boise Cascade plant at Marion, Oh.

Some of the functional advantages of the larger top-lift boxes were sacrificed with the quadrant boxes. The quadrants could not be top lifted and inserted two deep for reserve storage on rack shelves. They had to be transferred onto pallets for rack storage. The same was true when a palletload of four modules was positioned on the display floor. The two units in the back of the display had to be repositioned when the front two were empty in order to provide shopper access.

This led to the development at Basiloid of a unique two-wheel handcart. A simple slotted frame was welded to the front of the cart, and a lift blade dropped into the proper height slot. The blade was raised or lowered by means of a small hand winch and cable. To lift a carton, the handcart is positioned against it with the blade just below the top fold. The hand winch is then used to raise the blade under the top fold and to lift the carton. It is difficult to transport a 100 lb (45 kg) carton with a conventional two-wheel handcart. Considerable effort is required to get the handcart base plate under it, and then to pull the load back over the axle. It is equally difficult to deposit the load without it toppling over. The special top-lift handcart, however, makes it easy to do. The carton is lifted from the top, and no base plate is needed. Cartons weighing up to 500 pounds were easily lifted and transported with the top-lift cart. (Figure 13.2)

Demonstrations of the top-lift quadrant boxes and the unique handcart were impressive. Changes in the management organization at both General Foods and its warehouse store customer, however, delayed further progress with the project at the time, and with the retirement and transfers of key personnel, the display project was eventually dropped.

Basiloid continued the development of the handcart, and as others heard about it, as well as the experimentation with bulk displays of merchandise, inquiries were directed to Basiloid.

One of the first applications of top-lift handcarts was at York Corporation, a manufacturer of air-conditioning equipment at Norristown, Penn. Air conditioners are packed in large corrugated cartons that have top-lift folds on one side. The cartons are lifted and stored in clusters of two or three at a time by lift trucks with Basiloid blades. The handcarts are used to transport individual cartons to load and unload pickup trucks for local deliveries.

TOP-LIFT AUTOMATION

Another unique equipment development for top-lift packaging (Figure 13.3) took place in 1987. The Williamson Company, a manufacturer of heating and cooling equipment in Cincinnati, Oh., wished to reduce the packaging costs of top-lift cartons and

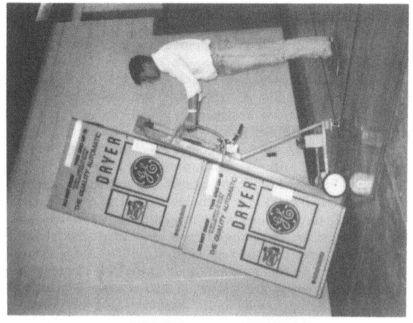

Figure 13.2 Top lift handcart photographs contributed by Basiloid Corporation, Elnora, Ind.

to find a more efficient way of applying perimeter straps to hold down the top folds. The problems were taken to the Liberty Nutro Company of Girard, Oh., a company that manufactures stretch-film wrapping machines. A packaging system called Stretch Pak resulted from the development effort (Figure 13.3).

The pack consists of four E-flute corrugated corner posts, a bottom tray, and top cover. An appliance is placed into the bottom tray and the four corner posts and top cover are installed around it. The posts are designed to interlock with the bottom tray and top cover. The partially packed appliance is then placed onto a stretch-film wrapper turntable and wrapped with a series of bands of stretch film. As the stretch film is being wrapped around the load, a cinching device closes the wide band of film to form a rope of film around the bottom tray, and again around the top cover perimeter. This firmly secures the tray and cover to the corner posts and provides a strong top lip for the insertion of top-lift blades.

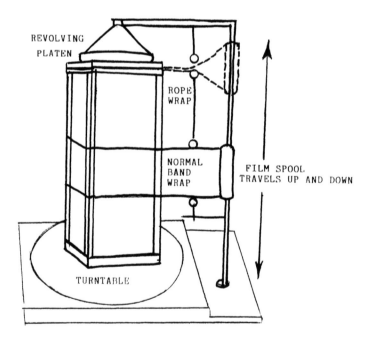

Figure 13.3 Top-lift container wrapper concept.

The method eliminates the hand taping or banding of the topfolds of top-lift cartons. Since so little packaging material is used in the process, packaging costs are minimal compared with conventional top-lift bulk cartons.

The case history of the development of top-lift packaging methods and equipment illustrates well how new bulk packaging concepts and ideas are initiated and evolve. A packaging method that may have been found to be impractical or unaffordable at one time may become a viable method at another time as new packaging materials and equipment are developed.

14

Case History 4: Wheeled Bulk Modules

Packaging for unitized handling must be functionally and economically suited to the particular logistical system. In other case histories discussed in this text, the projects were centered around the development of new and innovative bulk packaging. Materials-handling equipment and methods were tailored as necessary to packaging requirements.

The examples discussed in this chapter cover systems in which the process was reversed. The materials-handling methods and equipment received design priority, and the packaging was designed for cost efficient use of the materials-handling methods.

The urgent need for the rapid deployment of inventory in relatively short-haul, closed-loop systems focused attention on materials-handling methods in the systems. With closed-loop systems, the transport vehicles are usually owned or operated by the manufacturer of the goods. The goods are delivered to the buyer, and the vehicles are returned to the original shipping docks. It is thereby possible to utilize materials-handling devices, such as wheeled cages, in the deployment of inventory, since empty cages can be returned in the same vehicles.

Wheeled cages are commonly used for short-haul deliveries of many kinds of food products such as baked goods and dairy products. The tare weight and space occupied by the cages in the vehicles are an inherent part of the system, but there are offsetting benefits. The goods are never high stacked in warehouses or transport vehicles. The dimensions of cartons are not so critical, since they are contained within rigid metal cages. Relatively light packaging is, therefore, adequate, and secondary packaging such as shipping cases may not be required. Packaging costs are minimal, and there is less packaging waste to dispose of on the user end.

The cages make possible unitized handling without pallets and fork-lift trucks. They are pushed or pulled by hand into the vehicles at the shipping docks.

The comparative labor productivities for loading the same products into vehicles by hand stacking, by lift trucks and pallets, and by wheeled cages are approximately as follows:

> Move the pallet of products into the vehicle and hand stack the products — 1–3 tons/man hour.

> Enter vehicle and deposit loads by lift truck — 10–20 tons/man hour.

> Push wheeled cages into vehicle and secure them in place — 8–15 tons/man hour.

The technology for wheeled unitized systems is most advanced in Europe. Wheeled cages may seem primitive in comparison with the newer kinds of wheeled systems in Europe. The light packaging feature of cage systems may or may not be applicable. It depends upon the particular system.

The role of packaging in these advanced wheeled systems may be better understood if we review how and why such systems originated, and the major objectives in selecting the wheeled systems for specific applications.

A key objective in all the systems is the fast deployment of inventory, which requires the rapid loading and unloading of vehicles at the dock. Perishable products, such as fresh eggs or dairy products, in which there is a need to rapidly move the goods across docks and into air-conditioned areas, are good candidates for wheeled unitized systems.

The first large installation of a wheeled system was in the year 1966 for Bahlsen, a producer of packaged baked goods at Hannover, Germany. The logistical system required the daily dispatch of large numbers of truckload shipments of products to 50 regional depots throughout Germany. Rapid loading at the dock was critical. Hand pushing cages of goods into the lorries was much too slow as the shipping volumes expanded.

A pallet system was considered as an alternative. Cases of goods would be unitized on wooden pallets and warehoused. For outbound shipments, lift trucks would transport palletloads to the docks and transfer them onto lines of roller conveyors. The transport vehicles or lorries would have roller conveyor sections installed on their flooring. Powered rollers in the line would be used to move the entire truckload of unitized goods into the lorry within a couple minutes. The pallets would be locked in place and remain on the roller conveyor for the trip. Upon arrival at a regional depot the lorry would back to a raised dock, and the entire load would be pulled out by a cable and winch onto a continuous roller section on the dock. Empty pallets from the previous day's deliveries would then be rolled into the vehicle for return to the central plant.

The pallets/roller conveyor concept had several shortcomings. The main concern was the flexibility and speed of keeping the rollers in the marshalling areas on the docks loaded with the proper palletloads. A sizeable crew of lift trucks would be in continuous operation. Also, the palletloads would be stacked on the floors of the warehouse, and packaging would have to be strengthened to resist compression and bulging.

Finally, installing and maintaining the long lines of roller conveyors and the auxiliary equipment needed for the system would have been very costly.

Bahlsen retained a Swiss consultant, Hans R. Haldimann, to study the problems and to determine if a more flexible and economical alternative to a pallet/conveyor system was possible.

Mr. Haldimann's study and research led to the idea of a mobile platform consisting of a sheet of plywood to which four rigid mounted nylon casters are attached. The dimensions of the surface were 75×115 cm (29.5×45.2 in.), so that three could be fitted across the inside dimensions of a highway lorry. Rigid mounted casters are relatively inexpensive in comparison with swivel-type casters normally used on wheeled cages, and very little maintenance is required (Figure 14.1).

Rails for the wheeled platforms were installed throughout the production areas of the plant, in warehouse racks, in the marshalling areas on the docks, and inside the transport lorries. All rails were sloped to a 1.8% grade to move the wheeled loads by gravity throughout the system. Pavement in front of the docks is also sloped for transport vehicles to accommodate gravity flow loading and for the discharge of the wheeled loads. The surface of the rails on one side of the tracks is U shaped, which keeps all four wheels on the tracks and prevents derailing.

Unlike roller conveyors, the tracks on the docks are slightly recessed into the dock surface and do not obstruct cross traffic by personnel or mobile equipment when they are not in use loading or receiving wheeled loads. If it is necessary to remove a load from the rails, it can be easily moved manually on a flat surface area. A simple tug device, consisting of a long handle with two small wheels mounted in a frame on one end, facilitates handling. The wheeled end is positioned under the front of a platform, and when pushed down, the frame over the tug wheels lifts the wheels of the platform just enough to permit steering the load as it is pulled over a flat surface.

The height of each wheeled platform with the load is limited to 155 cm (61 in.) in order to fit into lorries equipped with two levels of rails.

(a)

(b)

Figure 14.1 (a) Integrated pallet with metal carrying frame on nylon wheels, (b) the wheeled base consists of nylon tread wheels mounted in metal frame. Contributed by Julius Minder, logistics consultant, Zurich, Switzerland.

Since packaged and cased goods are never stacked more than a few layers high on the wheeled platforms, the light packaging feature that is common to all wheeled cage systems applies, and packaging costs are minimal.

In order to control the rate of speed on the 1.8% inclines, Haldimann designed an ingenious frictionless braking mechanism. It consisted of a large rubber-treaded wheel mounted between the rails at about every 2 m (6–7 ft). The wheel turns on a hydraulic cylinder at a speed that is governed by the adjustment of the rate of flow of the hydraulic fluid. It essentially becomes a small hydraulic pump as it turns. These wheels are installed high enough to make contact with the underside of a wheeled platform descending along the rails. As contact is made, the wheeled load is braked to the maximum speed of the revolving hydraulic wheel.

The Bahlsen system required the development of different kinds of auxiliary equipment. A powered endless belt device was designed to transfer wheeled platforms between mobile stacking equipment, warehouse racks, and marshalling areas on the docks. The devices are mounted on the beds of stacker machines or serve as special lift-truck attachments. Mechanical stackers and destackers were designed to store and dispatch empty wheeled platforms.

Second-generation improvements in the technology included an improved method of controlling the rate of speed of wheeled platforms on the sloped rails. With this method the slopes are very slight, about 1/32 in. per linear foot (0.08 cm per 30 cm), which is barely visible to the eye. Static friction keeps the wheeled platforms from rolling down such a minor incline without a tug or push. Once underway, however, the loads continue to roll at a slow, even pace for short distances before rolling friction brings them to a stop.

Two kinds of mechanical devices are used to provide the initial propelling force. For light loads, a series of plastic ratchets are installed down the center of the rail lanes. As a wheeled platform contacts a ratchet in the line, the ratchet applies pressure against it to propel it forward. As the load passes over it in the line, the ratchet is automatically spring-charged to apply pressure against each succeeding wheeled load coming down the line.

For heavier loads the ratchets are replaced by simple air pistons to which cables with tug bars are attached. The air piston is charged by the force of the first wheeled load passing over it in the line. Upon discharge, it pulls the cable and tug bars to start succeeding loads to move down the line.

No braking devices are required with these methods. Loads, rolling down rack rails or into transport lorries, are allowed to impact. Cushioning devices are sometimes used to absorb the impacts of the heavier loads, but in most cases the impact is so slight that no further cushioning is needed.

A human-driven massive stacker machine, which is capable of transporting 12 wheeled pallets at a time to and from racks and to the load marshalling areas on the docks, was designed for the initial installation at the Bahlsen plant at Hannover, West Germany in 1966. In later years the machine was fully automated with computerized controls. Within recent years the Bahlsen system has been expanded to service over 100 regional depots from the central bakery at Hannover. Over 300,000 wheeled platforms were put into operation in the total system.

Sales demand does not always support the shipment of full unitloads of single items. The sales volumes of a number of items necessitated breaking down the full unitized loads and rebuilding mixed unitloads made up of layers of two or more different items. An innovative packaging/unitizing system was designed to speed the breakbulk and assembly operations.

The products to be assembled into mixed loads are unitized tier by tier on kraft paper slipsheets, usually four or five tiers per wheeled platform. Wheeled loads of the products are directed onto rails on either side of an assembly line. A mobile transfer machine operates laterally down the center of the assembly line aisle. A gripper bar, similar to that of a slipsheet push-pull attachment, is extended from this machine to grip the slipsheet lips and to pull one or more tiers of a particular product onto a wheeled platform carried aboard. When a unitload is fully assembled, the mixed load is delivered on its wheeled platform to a stacker machine that transports it to the marshalling area for outbound shipments.

The original Bahlsen system was installed in a time that predated the use of stretch film to overwrap and keep loads intact. Cases dimensioned to form into compact and stable unitload patterns were, therefore, especially critical to the system handling efficiency. Loads that bulge or lean were intolerable in the system.

Upon completion of the Bahlsen project, Haldimann continued to develop and market wheeled unitized systems. The wheeled devices were called Roll Pallettens in Germany, Pallettes Roulantes in France, Dollies in England, and Buggies in the United States.

To accommodate users who wanted to keep the flexibility of wooden pallets in their systems, Haldimann designed a special wheeled device to carry a standard European pallet. It consisted of a lightweight metal frame to which 4-in. (102-mm) nylon casters were attached. The device with casters weighs just 14 lb (6.4 kg). The wheeled pallet system was marketed under the trade name, Rollax, and a number of installations were made throughout Europe.

One of the most imaginative installations of Haldimann's wheeled systems was built in 1973 for a brewery at Basel, Switzerland (Figure 14.2). It involves the use of wheeled platforms with a metal grid surface. Packaging for this project is reused. Reusable beer bottles are encased in thermoformed plastic crates, which are unitized automatically on the wheeled platforms, and which are put through warehousing, staging, and dock loading operations on sloped rails. The return of the empty bottles and crates is integrated into the total system.

The project was initiated by the Warteck Brewery at Basel to provide warehouse storage and dock space at the plant site. Space for warehousing was limited in the existing building, and there was no space available in the crowded industrial area to build a remote warehouse and docking facility. A decision was made to excavate a gravel pit that lay under the brewery building and to build a subterranean storage facility.

Four levels of sloped rails were built in racks in the subterranean chamber. A double-lane roadway was constructed at street level over the warehouse. It was sloped on either side to accommodate the gravity flow of wheeled loads.

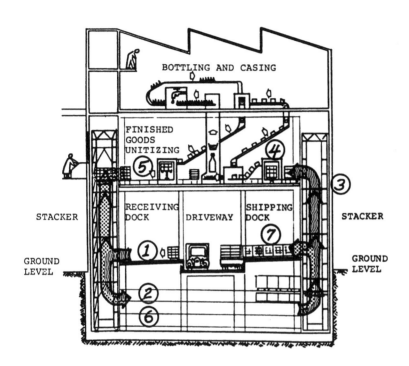

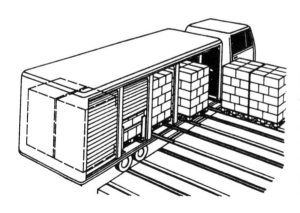

Figure 14.2 Warteck Brewery at Basel, Switzerland. A fully integrated system based on wheeled pallets. Glass bottles, plastic cases, and wheeled pallet bases are recovered and reused. Courtesy of Julius Minder.

The wheeled loads are transferred from a third floor unitizing area to the underground racks by means of an automated stacker crane that delivers up to six loads at a time. A twin stacker on the opposite side carries outbound loads to dock staging areas.

Route delivery vehicles, returning with loads of empty plastic crates with beer bottles, pull up alongside the sloped receiving dock. A side gate is removed, and a short dock plate is put in place to connect the deck of the vehicle and the dock. The driver releases a holding bar, and the wheeled platforms carrying unitloads of crates and bottles move by gravity onto the sloped dock area. From there the loads flow onto the stacker crane for transport to the top level of the building, where they are washed and sterilized. The clean bottles are subsequently refilled and automatically loaded again into the plastic crates. The crates, in turn, are automatically unitized onto the wheeled platforms for dispatch by stacker crane to the underground warehouse.

The wheeled loads remain on the sloped rails in the underground storage area until ready for dispatch. A twin stacker crane on the opposite side of the warehouse is used to transport them to the staging rails on the outbound shipping docks. Outbound vehicles are loaded from the side and dispatched rapidly.

Futuristic handling systems such as these may come into wider use in the future if the need for rapid deployment of inventory continues. Totally reusable packaging as found in the system at Basel may be the exception. Most wheeled systems make possible the use of light, disposable packaging that is relatively inexpensive.

15

Case History 5: Bulk Goods Modules

The conventional, small corrugated case of merchandise offers marketing flexibility in the complex distribution systems for packaged goods. Small buyers may order just one or two cases of a product, while large buyers may order larger numbers. The economy of scale benefits that are applicable to bulk containers are sacrificed for this system flexibility.

The 1980s saw a trend toward larger and larger retail outlets for general merchandise. Wholesale club stores, warehouse stores, and hypermarkets were growing and expanding at a tremendous rate. Such stores buy the manufacturers' goods in palletload quantities. The pallets are set out on the retail floor, and cases are cut open for access to the goods by the consumer. The idea is to eliminate the labor-intensive activities of opening cases, hand transferring goods onto retail shelves, or cutting down the cases and stacking them into displays at the ends of retail aisles.

Many packaging experiments have been made to improve the mass merchandising operations. One large warehouse-type store used labor crews to open cases and stock the product into pallet-size wire mesh tote boxes for delivery to the retail floors by lift trucks. The method eliminated the accumulation of tons of corrugated waste materials on the sales floors, but the multiple handling operations offset much of the gain in productivity. Another large manufacturer packaged his products in special display cases that were convenient to build into mass displays on the sales floors. The top-lift box concept described in Chapter 13 was another attempt to develop a giant box for the display of goods on the sales floors of giant supermarkets and warehouse stores.

CASE HISTORY—T-BOX PREBUILT STORE DISPLAYS

The story of the T-Box provides an interesting case history in the search for an improved means of packaging and distributing packaged food products. It is also a classic example of the frustrations inherent in the introduction of a new and innovative method into an established system built upon and grown accustomed to the more primitive methods of handling and shipping merchandise.

The T-Box, an atypical single-sided container, in which the packed goods ride on the outside, is more than a box. It is a unitized shipping system for open-front displays of goods packed at manufacturing plants and shipped through distribution channels to retail stores. It provides retailers with an instant display, a time-saving alternative to the labor-intensive practice of opening many small cases of products and manually building product displays piece by piece on the sales floors (Figures 15.1 and 15.2).

The T-Box method employs a relatively small amount of packaging material to protect the open fronts and tops of displays during shipment. A T-shaped section of paperboard is positioned upright through the center of a unitload of four displays. Two displays are placed into a tray base on either side of the upright section with their open sides pressed tightly against it. Wing flap extensions at the top, which fold down over the open tops, function both as a protective cover and a means of locking the four

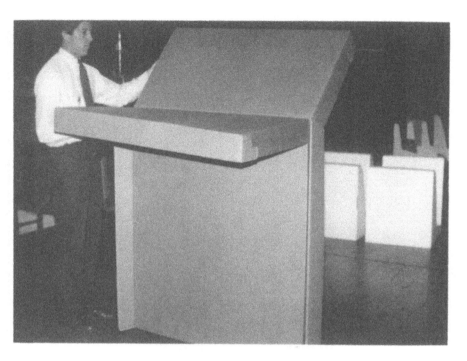

Figure 15.1   The T-box.

displays together as a unitload. Rigid backs, corners, and tapered sides of the displays, in combination with the vertical T section, provide product protection plus top-to-bottom compression strength throughout the load. This enables high stacking without crushing the displays at the bottom of a warehouse stack.

Each display is a quarter-size module of the standard pallet load, approximately 24 in. wide, 20 in. deep, and 45 in. high (610 × 508 × 1143 mm). The quadrants are tall and wide enough for good visibility on sales floors, and the depth puts all retail packages within convenient reach of shoppers. The height limit was necessary to ensure that the displays can be double stacked for the effective use of space in trailers.

Experimentation with different kinds of packages in the displays determined that most products should be packed first into trays, and the trays are then packed into the quadrant. Although

Figure 15.2   Loading displays into T-box module.

trays add to the packaging costs, and introduce a secondary pack operation, they serve two functions that justify their cost. First, they prevent tall packages from toppling out the open fronts of the displays. Secondly, trays provide a convenient means of transferring residual packages from the bottoms of displays nearing depletion to the tops of full displays or onto store shelves.

For the initial application, it was difficult to make a decision on the size of the display and the quantity of goods it would carry. The larger the display, the greater would be the cost benefits to the production, handling, and shipping operations. On the other hand, smaller displays would enable a greater number of stores to participate in the display program. Therefore, a compromise resulted that provided for a minimum-size display that could be ordered as a single sales unit for the smaller stores. Multiples of it could be combined to form larger displays for larger stores.

Design concessions were also made in order to produce and distribute the displays through the same channels as conventional cases. Functional and cost objectives were established for each step of the manufacturing and distribution process, including:

Packaging: Costs could not exceed the packaging costs of an equivalent number of conventional cases, and tare weight should not exceed that of the conventional cases.

Production: Displays must be adaptable to machine packing and assembled with conventional case packing, conveying, and unitizing equipment.

Unitized shipping: Displays, like small cases, must be dimensioned to be compatible with the standard 48×40 in. pallet. Slipsheets were to be used for long-haul truckload transport from plants to regional distribution terminals, with the option of transferring the slipsheet loads onto pallets for the final short-haul delivery to the stores.

Terminals: Unitload breakdown, selection, and the assembly of displays for retail store orders must be compatible with conventional case-handling operations and must not exceed small-case-handling costs per net ton.

Transportation: Costs per net ton shipped should not exceed that of the conventional cases. The height of displays will be critical to the effective cube utilization of transport equipment.

Store handling: Individual displays must be capable of being handled by two-wheel handcarts that are commonly used to transport cases from store back rooms to the sales floors.

Sales: The manufactured displays must be as attractive on the sales floors, and as convenient to shoppers, as the displays that are built from small cases.

It was apparent early in the design of the T-Box that it could never satisfy the functional and cost objectives of all the independent areas of activity along the logistical chain, since some were in conflict with others. Compromises had to be made, and in some instances this meant an increase in costs in one or more areas to achieve the overall system cost efficiency projected for the T-Box.

The first T-Box shipping tests involved retail packages 6 in. wide, 2 in. deep, and 7.5 in. tall ($152 \times 51 \times 190$ mm). They adapted well to targeted display dimensions in a configuration of 12 deep per row, four rows across, and six tiers high. This was the equivalent of 10 conventional cases of the product. The gross weight of each display was light and easy to handle by a two-wheel handcart. This permitted palletless handling and shipping throughout the system, thereby saving the costs of pallets. T-Box loads on slipsheets could be handled by lift trucks at plants and regional warehouses. Upon arrival at the stores, the T-Boxes could be opened in the transport vehicle, and individual displays transported to sales floors by handcarts. The shipping and handling tests of prototypes demonstrated the feasibility and potential cost benefits of the method.

The main obstacle to the expanded use of the T-Box method was the quantity of goods necessary to make a cost-effective manufactured display. Retail stores prudently keep inventories as low as possible, and prefer the flexibility of building bulk displays of merchandise, small or large, in order to be consistent with sales. Furthermore, the design modifications that had to be made to gain acceptance in as many functional areas as possible, within the existing system, added to costs and tended to erode the ultimate cost and productivity gains that could have been possible with the method.

If the trend toward mass marketing of consumer goods continues in the future, there should be opportunities for packaging people to develop similar creative bulk goods modules to eliminate the labor-intensive operations inherent in the small conventional cases.

In the meantime, vast opportunities exist for improving the productivity of goods in process that are currently shipped in small cases. Some examples follow.

A Canadian manufacturer of a rice product purchased pouches of a sauce mix from an outside supplier. The sauce pouches were packed in small cases and the cases shipped on standard wooden pallets. The pallets of cases were set near the production line, opened, and the pouches transferred by hand onto a conveyor line, where they were fed into cartons of rice coming down the line. The production operation, while infrequent, produced a substantial volume in a short time. Large piles of empty corrugated cases from the pouches accumulated and had to be continually stacked and baled for trash disposal.

The production operations people were offered the alternative of receiving the pouches in large, single paperboard boxes mounted on slipsheets. The boxes were 48×40×32 in. (1219×1016×813 mm) and weighed several hundred pounds. When empty, they collapsed into a returnable package just 6 in. (153 mm) high. To accommodate loading, a number of the pouches were loaded into a light plastic bag, and the bags were then dropped into a bulk box. At the receiving plant, the slipsheet boxes were transferred onto plant pallets and moved by a lift truck to the production line. There, the boxes were positioned onto a tilt stand. Production line workers would pull a plastic bag from the bulk box and dump the pouches from it onto the conveyor line. The light plastic bags were accumulated in a previously emptied box, and later discarded as trash. The bulk boxes were collapsed and stacked on a pallet for return to the supplier to reuse.

Another example of the integration of a bulk method into a production line operation was the packing of stacks of nested plastic bowls into a large bulk container. The existing method was to stack the bowls about 12 in. high and to pack them into corrugated cases. The cases were unitized on pallets, moved to the shipping dock, and hand stacked off the pallets into a semitrailer for shipment. The cases were not shipped on the pallets due to the transportation cost penalties. The product was bulky and lightweight. It was important to get as many cases as possible in the trailer in order to minimize the costs per net ton shipped. The costs of hand stacking the light cases were less than the transportation penalties that would have resulted if the shipment were on pallets.

An alternative developed by the packaging/distribution department of the buyer was the use of a large corrugated bulk box with a slipsheet base. The box, 48×40×30 in. deep (1219×1016× 762 mm deep), had to be integrated into the production line operations at three supplier plants and the buyer's plant. The boxes were too deep for conveniently hand packing tubes of nested bowls inside. Positioning the bulk boxes on tilted work tables helped, but did not solve the problem. A decision was made to preload the long tubes of bowls into a light plastic sleeve and then to load the sleeves into the box. A simple machine was developed on which a plastic sleeve was inflated by air for the convenience of feeding in the tube of nested bowls.

The filled boxes were capped and shipped three high in stacks on slipsheets to gain the equivalent load density of the hand-stacked small case method.

The boxes were transferred onto pallets for the convenience of handling at the receiving location. At the production line, the large bulk boxes were set onto a platform that elevated them by means of a scissor-lift hydraulic system to the packaging machine operators. The operators would remove the box caps, pull out a tube of bowls, and place it onto the packaging machine feed hopper lip. The plastic bag was withdrawn from the tube and discarded into an empty box for trash disposal. When the large boxes were empty, they were collapsed and stacked for return and reuse. While the method resulted in an incremental cost for the disposable plastic sleeve, the overall system costs were reduced. The plastic sleeves contributed to improved sanitation practices in handling the bowls.

The case examples illustrate the opportunities for the use of bulk containers for in-process goods and materials. Projects of this kind are initiated in different ways. The supplier may develop the bulk packaging and shipping method as a service to his customers. In other cases the user may develop the concept and request the supplier's cooperation in converting to the bulk system. Sometimes a bulk container manufacturer may develop and patent an innovative system and conduct a marketing program to introduce the system to potential users. The proximity of the supplier location to the user location often discourages the introduction of new and innovative bulk container systems. A supplier who is located thousands of miles from the receiving plant may not know if the receiver is interested in alternative methods and consequently packs and ships the product in conventional cases unless requested to use an alternative method. The user, on the other hand, may not be aware of alternatives that may be possible, and without question adapts the receiving plant operations to handle the traditional-sized cases.

Initially, the integrated packaging design approach requires a management commitment to the study of alternative methods of packaging and shipping the goods and materials. In making this commitment someone must be assigned to carry out the research and study necessary to identify the opportunities and to develop the alternative methods. The packaging specialist is in a good position to handle this kind of assignment, since packaging design is critical to the success of the total system.

16

Case History 6: Strap-Suspended Modules

The case history of strap methods of packaging provokes an interesting contrast to traditional methods of packaging and handling items applicable to this method. It also provides insight into the innovative bulk packaging design process, which in this case included the design of the tools and machines needed to fabricate and assemble the unique box.

It illustrates well the fact that innovation in bulk packaging is usually the result of building and expanding upon a concept, as opposed to having its origin as a preconceived idea in the mind of the packaging designer. The concept in this case history evolved out of the search for a cost-effective means of packaging and shipping fragile automotive subassembly components. It began with the use of straps as a built-in materials handling device for an exceptionally large corrugated box and triggered the idea of the use of straps as an internal system within a large box.

The background to this development will familiarize the packaging designer with the possibilities that exist for departure from the traditional packaging design approach.

The project originated at General Motors Cadillac Division in 1982. A decision was made to enter the ultra-luxury sports car market. This began a worldwide search for the best quality parts and body styling. The search ended with a contract with Industrie Pininfarina, an Italian firm famed for its body designs of cars such as Ferrari, Lancia Aurelia, and Alfa Romeo.

The manufacture of the car, the Allanté, was complex in that the engineering design and the production of the underbodies, engines, and powertrains could best be done in Detroit. The installation of the body and accessories, on the other hand, could best be done in the Pininfarina plant at Turin, Italy.

Production plans for that reason called for the manufacture of underbody components, cowlings, and floor pans at Detroit. They would then be packaged and shipped to Turin by air freight for the installation of bodies and accessories. The partially completed cars were to be returned air freight to Detroit and put back on the line for installation of engines, powertrains, and final assembly.

Shipment by air required special attention to the packaging and materials handling methods. While important to all modes of transport, two key objectives are especially critical to air freight systems:

> Full utilization of space in the cargo holds to achieve the highest density load possible. The more parts per shipment, the fewer the number of flights required.

> Minimum tare weight of packaging and materials-handling devices in order to gain the highest net tonnage possible. Fuel consumption, a key economic factor in air transport, is related directly to gross tonnage. Consequently, the less the tare weight, the less the fuel consumption per part shipped with more favorable impact on total freight costs.

The development of packaging to transport the fragile underbody cowlings and floor pans was particularly critical. Wooden crating would normally be used for such freight, however, it would increase the tare weight substantially and impact air freight costs unfavorably. The use of corrugated paperboard could reduce tare weight, but raised concerns for adequate compression resistance. Furthermore, a corrugated case large enough to house the underbody of a car would exceed the size of any commercially available paperboard box. Such a large paperboard box could be expected to be difficult to set up and pack. Materials handling would also be a problem. A huge wooden pallet could be placed beneath it to accommodate fork lift truck handling, but that would only add further to tare weight and occupy space that would limit the number of boxes that could be gotten into the hold of the aircraft.

The packaging design problems were taken to Julius B. Kupersmit, the designer of the palletless K boxes that were handled by means of a built-in slipsheet. Subsequently, a consulting contract was signed with Mr. Kupersmit's firm, Containair Systems Corporation, to develop prototype boxes for the air freight shipments of parts from Detroit to Turin, Italy.

Mr. Kupersmit's research included visits to the manufacturing and assembly line operations at both Detroit and Turin and inspection of the cargo bays of the Boeing 747 jet freighters that would be used in the transatlantic shuttle. A step-by-step flow chart was put together to trace the route the containers would make from the General Motors Cadillac plant near Detroit to the metropolitan airport, and from Turin's Caselle airport to the Pininfarina plant near Turin.

Samples of the underbodies and the parts to be shipped with them were sent to the Containair plant near JFK airport in New York for use in the building of prototype containers. A team of General Motors people assigned to the project assisted in the development of packing procedures. It was determined that two underbodies could be nested together to achieve excellent pack density. A large corrugated box was built around the two nested parts (Figures 16.1–16.3).

Although corrugated material is relatively lightweight compared with wood, it too can add substantially to tare weight, depending upon the kind and thickness selected. To minimize tare

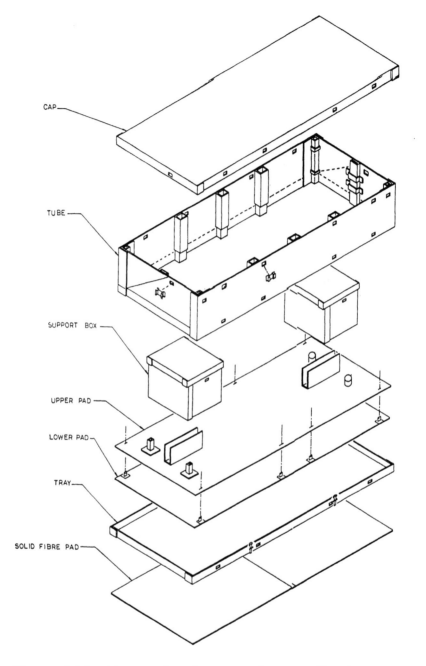

CAP

TUBE

SUPPORT BOX

UPPER PAD

LOWER PAD

TRAY

SOLID FIBRE PAD

Figure 16.1 Air cargo box for automotive parts. Drawing courtesy Containair Systems Corporation, Springfield Gardens, N.Y.

198

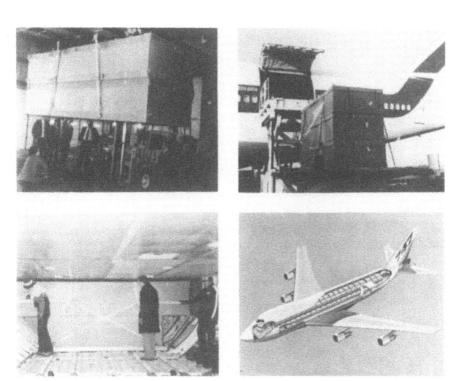

Figure 16.2 Allanté box. Photos courtesy, Cadillac Motor Car Division, General Motors Corporation, Detroit, Mich.

weight, Kupersmit used relatively thin, double-wall corrugated board. Borrowing upon the design of his K boxes, he provided scorelines in the end walls in order to collapse them onto the base. A removable panel on one of the long sides and a removable top cover provided access for the loading and unloading of the parts. Tubes formed out of the double-wall corrugated board were used to strengthen the walls and corners of the big box. This provided top-to-bottom compression resistance so that the boxes could be stacked up to three high.

Precise dimensions and specifications for the paperboard parts was critical to the container design. The design called for scores and slots to be incorporated with the manufacture of large,

Figure 16.3 Allanté parts box. Photos courtesy of Cadillac Motor Car Division, General Motors Corporation, Detroit, Mich.

flat corrugated sheets. The locations of the corrugated tubes and other pieces were imprinted on the large sheets during the production run.

The flat sheets had to be formed into the various components, some 85 pieces in all, then stapled, glued, and assembled. The design of the tools, working tables, and machines necessary to the assembly process were specified or designed by Mr. Kupersmit. The secondary assembly operations represented a key cost in the manufacture of the box and depended a great deal on the design of components that adapted well to the use of simple tools and machines.

Miscellaneous parts were packed into two large pallet-size corrugated boxes, 34×31×28 in. (864×787×711 mm). It was found that there would be room to fit the two boxes in openings in the underbodies within the main box and thereby not only gain excellent storage density for the miscellaneous parts, but also utilize the two boxes inside to contribute top-to-bottom compression resistance inside the main container. The boxes were accordingly called support boxes.

The method of securing the support boxes to the floor of the main box was critical. Pallets could have been glued in place and the bottom flaps of the inner boxes nailed onto them. The pallets, however, would add too much to the tare weight and would reduce the cubic capacity. The solution was the use of a section of light, double-wall corrugated board, 26×29 in. (660×737 mm), that contained a scored 4 in. (102 mm) center strip and 12 in. (305 mm) flaps on either side.

The center strip was glued and stapled to the precise position, and the flaps folded up and inserted through a slot across the bottom flaps of the support box. The flaps were then folded down on the inside of the support box to interlock with its bottom flaps. With this method, the support boxes were anchored securely into place inside the main box.

The fully assembled box measured 14.5 ft (4.42 m) long and 6 ft (1.83 m) wide. When collapsed with the cap in place, the height was just 7½ in. (19 cm). The collapsibility of such large boxes is important to achieve as efficient a shipping density as possible. The greater the number of empty boxes per truckload, the less the cost per unit shipped. In this case, the empty boxes were stacked 12 high and strapped as a bundle with cross boards at the bottom to permit the entry of lift-truck forks. High cube trailers with a doorway height of 9 ft (2.74 m) were required for shipment of the 12-high stacks from the manufacturing plant in New York to Detroit. The unitloads of 12 boxes were not dimensioned for the effective use of the highway trailer cube. Only two to three stacks could be shipped in a truckload dependent upon the length of the trailer available. The length of the largest highway trailers available on the shipping route was 48 ft (14.6 m). The distance from the manufacturing point to the point of use was, therefore, another consideration in the production of large boxes of this kind. In this case, it was convenient to manufacture the boxes in the designer's plant in New York. If the volume anticipated for the project increased, as expected in the future, it was intended to relocate the box assembly operations closer to the point of use in Detroit in order to minimize the empty box shipping cost penalties.

The boxes, each packed with two underbodies and miscellaneous parts, measured 174×72×31 in. (442×183×79 cm). This size allowed the boxes to be stacked three high and fitted into the upper cargo holds of the Boeing 747 freighters.

A decision had to be made on the materials-handling method for the heavy boxes, which weighed approximately 1500 lbs (680 kg). The traditional approach would have been to strap the big boxes to wooden pallets and to use fork-lift trucks to stack and handle them. That, however, would have limited the number that could be shipped. The added height of the pallets would have limited the stacks to two high in the upper hold, and the tare weight of the pallet would further increase shipping costs. The packaging designer was faced with developing a suitable palletless method of handling.

The initial thinking was to glue large, heavy-duty slipsheets on the bottoms of the boxes and to handle them with push-pull attachments on the lift trucks. A prototype box with a slipsheet attached was built, loaded, and tested. Although the slipsheet method proved to be feasible for handling and stacking the boxes, an unexpected problem surfaced during the positioning of them into the cargo bays. The base of the bottom box in the stack had to be fitted into a 4 in. (10.16 cm) metal frame that held the loads in place inflight. It was necessary to lower the load down to fit inside the frame, and that could not be done effectively with the slipsheet method of handling.

The idea of strap handling evolved following consideration of all known palletless alternatives. Two nylon-web straps, each 80 in. long (2032 mm) by 1 3/4 in. (45 mm) wide, rated to 6000 lb (2722 kg) tensile strength, are sandwiched between a solid fiber sheet and a corrugated tray on the bottom of the box and held in place by glue. The ends of each strap extend 3 in. out from either side of the box, and a steel D-shape ring is fastened to the end of each strap.

A simple H-shaped steel rod frame was designed to fit onto the forks of a lift truck. Four web straps with D-ring latches on the ends extend from this frame device. To lift a box, or a stack of two to three boxes, the lift truck positions the H frame over the top of the boxes. The straps extending from the frame are latched to the D rings of each strap end of the box. As the lift truck mast is elevated, the boxes are lifted by the straps and are capable of being stacked and lowered down into the frames on the cargo deck.

The strap-suspension materials-handling method also adapted well to handling the boxes on the production lines. The H frame could be attached to an overhead crane, as well as a lift truck, and handled in that manner.

The strap-suspension method required a strong box bottom to avoid collapse under the pressure of the load when lifting. This was accomplished through a series of laminated corrugated sheets, with the direction of corrugations reversed in each layer, and the use of one solid fiber sheet. The reinforced bottom, in addition to making the box suitable for strap handling, facilitated the transport of the boxes on roller beds in the shuttle vehicles that transferred them from the plant at Detroit to the Detroit Metropolitan Airport for loading into the 747 freighters.

The 14.5 ft wide (4.4 m) boxes were tailored to fit across the inside width of the plane and to stack three high in the upper holds. The lower cargo hold of the plane, due to the curvature of the belly, permitted only one box that wide to fit across the hold. To avoid waste space in the aircraft, a special parts box dimensioned 114 in. (2896 mm) was designed to fit the curved base and to support one of the larger boxes on its top. Certain bulk parts were packed in this box. The unique boxes made possible the shipment of the underbodies and miscellaneous parts for 165 cars.

VERTICAL LAYERING PACKAGING CONCEPT

The success of the Allanté bulk packaging system led to the harness project. This involved finding an improved method of packing and shipping electrical harnesses from a manufacturing location in Mexico to General Motors' Cadillac plant near Detroit, Mich.

Harnesses are large assemblies in which multicolor coded wires are encased in a plastic housing. Harnesses are installed in cars on the assembly line, and the ends of the wires connected to a myriad of instruments and lights. The existing method was to pack each harness into a corrugated sleeve, and then to pack four sleeves into a large corrugated shipping case. Four cases were unitized and strapped onto a cheap, one-trip wooden pallet.

The method had many problems. The lightweight, bulky cases on pallets resulted in very poor payloads for truckload shipments from Mexico to Detroit. Transit damage to the fragile plastic parts was quite high, and many harnesses had to be rejected at Detroit. Finally, after unpacking, the sleeves and cases had to be accumulated and baled for disposal, and the wooden pallets had to be collected and disposed of as scrap wood.

Kupersmit's research included visits to the production plant in Mexico as well as to the assembly line of the Detroit plant to identify functional and cost areas for potential improvement. Out of this preliminary research came a list of key design objectives, which included:

Improve the shipping load density to get more harnesses per truckload shipped and to reduce transportation costs per piece on the long haul from Mexico to Detroit.

Reduce labor time and costs to pack each harness in a sleeve, then pack sleeves in a case, and finally unitize the cases onto a pallet.

Design packaging that will reduce the vulnerability of the harnesses to damage in transit.

Reduce the large amount of waste packaging materials that must be disposed for each harness shipped.

A supply of rejected harnesses was subsequently shipped to Kupersmit's plant in New York for use in building prototypes. The design objectives themselves provided guidelines for the design of the improved packaging system. The problem of waste disposal indicated a need for a reusable container. In order to reduce packing and unpacking costs, the container must be designed to be fast and simple to set up for packing as well as to collapse easily following unpacking.

Kupersmit decided to design a large container similar to the K box, the sides of which fold along score lines to collapse into its base dimensions. The outside dimensions of this box would be based upon the width of the harnesses and the need to effectively utilize the interior space of the transport trailer. The ultimate packed module turned out to be $58 \times 45 \times 50$ in. ($1473 \times 1143 \times 1270$ mm). The collapsed height when empty was 10 in. (254 mm). The modules would be capable of being shipped in a configuration of two high, two across, and nine deep each row in a high-cube, 45 ft long (13.7 m) highway trailer.

The next step was to find a convenient way to pack the harnesses into the large box and, likewise, to find an efficient way to unpack them. The traditional method would be to pack each harness in a protective sleeve or wrapping, and then pack as many as possible in the box. Substantial amounts of dunnage or cushioning material would be used to absorb the shock and vibrations between harnesses stacked inside the box and, thereby, to control damage in transit.

The traditional approach, unfortunately, would result in considerable waste materials. A search for a means to pack the harnesses without the need for large amounts of dunnage waste was initiated. It was noted that the mass of wires that protruded out from the plastic casings of the harnesses was actually suitable protective material itself. It would be ineffective if the harnesses were piled one on top of the other in the box, which would transfer the pressure of top pieces down onto those at the bottom.

The strap concept then came to mind. That is, suspend a series of plastic web straps from rods across the top of the box and attach the harnesses to the straps. The weight of each harness would then be independently suspended on the straps, and the bundle of wires that protrude from and surround each plastic piece would be adequate to stabilize the entire load in place.

Subsequent experimentation established that five harnesses could be suspended in one file, and a total of nine files deep could be fitted into the box, for a total of 45 harnesses for each box. This provided a pack density more than double that of the old method (Figure 16.4).

A subcontractor was brought into the project to manufacture the metal-rod internal structure of the box. To mount the rods, a section of heavy corrugated board was carried in the base of each box. This piece was mounted upright inside the outer walls of the box. U-shaped metal rods were clamped over the tops of the corrugated piece on each of the 45 in. sides. These rods served as tracks on which the crossbars were fitted to slide back and forth. A series of web straps extended from each crossbar. The longest straps contained the harness on the bottom of the file, and the shortest the harness at the top.

When the box was erected for use, the first rod was picked up from the bottom of the container and placed onto the cross bar. As each file of five harnesses was latched onto the straps, the rod was simply pushed back one file position and another one set into place.

The container design included a removable front panel and top cover for access to pack and unpack the harnesses. Unlike the original K boxes, the harness box was not equipped with a slipsheet for palletless handling. Instead, each container was permanently secured to a reusable wooden pallet. While the use of a slipsheet would have further increased the load density and reduced tare weight, the use of a pallet made the project more expedient to implement. The key reason for using pallets was that both shipping and receiving locations, as well as certain intermediate transfer points, were all equipped to handle pallets, but none had push-pull equipment for handling slipsheets. The transition to slipsheets would depend upon the potential to be gained by the elimination of pallet bases and the installation of push-pull attachments on the lift vehicles at a later date.

The first test shipments proved the method to be very effective. Damage in transit was virtually eliminated, and the integration of the loading and unloading operations on the production

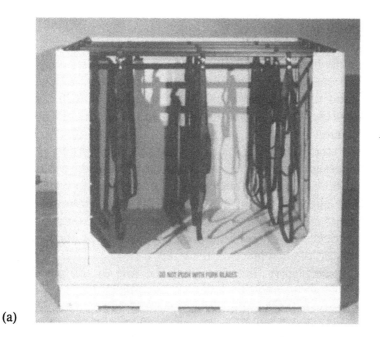

(a)

(b)

Figure 16.4 (a) Harness box set up for loading. (b) Harness box loaded. Photographs *courtesy* of Containair Systems Corporation, New York, N.Y.

and assembly lines substantially reduced handling costs. Following unloading, the containers could be collapsed on their pallet bases, stacked, and then shipped back to the supplier plant in Mexico. The inconvenience and costs of waste disposal of the old method were eliminated.

Kupersmit filed for a patent on the strap suspension system, which he designated the Vertical Layering Packaging System. It appeared to be a concept that could be introduced to pack many kinds of fragile appliances, accessories, or tools. However, some might require an overwrap to protect them from horizontal contact in a large pack. The key to the success of this kind of packaging is the suspension of each piece on the strap.

This case history illustrates that the role of the packaging designer may go beyond the design of containers that conform to the manufacturing constraints of corrugated-paperboard production-line machinery. While every effort was made to adapt the design to the in-line production of the flat sheets of paperboard, it was necessary to employ a secondary operation for final assembly of the parts. The skill of the packaging designer to design simple tools and machines to facilitate the secondary manufacturing process is critical to total manufacturing costs. Such skills are especially useful for the design of bulk transport containers.

17

Case History 7: Fabric Bulk Containers

Bags represent a relatively low-cost method of packaging for materials that are adaptable to bagging. The ratio of tare weight to the net weight of the product is very small, and the collapsibility of bags simplifies empty bag shipment and disposal after use.

Bags for retail use must necessarily be sized for the convenience of handling by the consumer – 10–25 lb (4.5–11.3 kg). Bags for industrial use are, likewise, sized for manual handling, however, the weight of many bags of industrial materials range from 50 to as much as 154 lb (23–70 kg). The bags may be mechanically filled and unitized on pallets or slipsheets for shipping, but at the end of the logistical chain they must be handled individually. Manual bag handling is slow and tiring. Back injuries are common among workers who must consistently lift and move the heavy bags.

In the late 1960s, large bulk bags and their mechanical handling systems began to appear as an alternative packaging method for conventional bags of industrial products. They were classified as IBCs for intermediate bulk containers. An early application of the bulk bags was for the transport of agricultural products. The bags were used to load the hoppers of crop duster aircraft or were suspended from tractors to dispense fertilizers directly onto the fields.

In 1982, the Japan Packaging Institute made a worldwide survey to determine where and how bulk containers made of fabrics were being used. Twenty packaging organizations in 15 countries responded. It was found that millions of fabric bulk containers were used annually to transport powder and granular materials of many kinds. The weight of the loads carried ranged from 500–1000 kg (1102–2204 lb). Most of the bulk bags were made of woven polypropylene fabric with nylon or polyester strapping. The bags were filled from the top, and the contents were discharged by gravity through openings or spouts at the bottom. In the case of one-trip bags, they are usually cut open at the bottom to discharge the contents.

Packaging people should be well versed in the technology of bulk fabric bags and the mechanical systems used for filling and discharging their contents. They should be aware of the kinds of materials that adapt well to bulk fabric bag handling and those that do not. They should also be aware of the pitfalls, as well as the benefits, and the economics of bulk bags in comparison with non-containerized or full bulk handling. That will enable them to advise management on the best packaging system for a particular product and to assume leadership in the implementation of the total systems.

Hundreds of applications of bulk bags could be used as a case history example. The one selected took place in 1985 in Sweden. It involved a test shipment of large bulk bags of coffee beans. It is used as a case history example on the design and application of bulk bags, since it illustrates the difficulties of introducing bulk bags as an alternative packaging system, and also the kinds of questions that must be addressed in order to bring credibility to the proposed system. In this example, decision making on the bulk bag system was hampered first by the changes taking place in transportation methods, and later by the proponents of alternative systems.

Green coffee beans have traditionally been shipped in burlap bags, weighing 132–154 lb (60–70 kg). Until the 1970s, the bags were hand piled in the holds of breakbulk ships for transport from supply sources. Upon arrival at the ports of receiving countries, stevedores piled the bags into cargo nets and hoisted them to dockside, where dockworkers sorted, weighed, and hand loaded them onto pallets for storage or for delivery to the roasting plants.

During the 1970s, the concept of slings was introduced to reduce system costs. Relatively low-cost labor at the ports of coffee-producing countries hand loaded 12–16 sacks into a sling made of polyester straps and prestaged them for loading onto ships. Cranes and hooks were used to lift the sling loads aboard and to lower them down into the holds. The slings were unloaded at receiving ports, two or more at a time, by ship cranes. The sling method enabled faster loading and unloading of ships at the ports, which reduced the turnaround time and improved the utilization of the vessels.

The opening of the bags and the dumping of the beans to process, however, continued to be a labor-intensive and backbreaking job at the roasting plants. By the mid 1970s, bulk fabric bags were already in wide use for other kinds of materials. The use of bulk bags for coffee bean imports seemed to make a lot of sense. Bulk bags are handled by straps, and strap-handling equipment was already in place for slingloads. The filling and discharging equipment for the bulk bag systems was relatively inexpensive. A major concern of green coffee buyers at the time was the expected resistance of coffee-producing countries to the potential economic impact on the burlap bag industry in those countries.

The distribution development area of General Foods subsequently undertook an experimental project in 1977 to determine if bulk bags could be manufactured of burlap material. Prototypes of a hybrid bag made of burlap fabric, and strengthened by bands of polyester strapping, were produced at Bag Corporation labs at Savoy, Tex. A series of tests made at the lab in June that year demonstrated that the burlap bulk bags could contain and safely transport a ton of green coffee beans, or the equivalent of 15 conventional burlap sacks of beans.

The timing for the introduction of a new packaging system, however, was not right. Containerized ocean shipping was expanding throughout the world and replacing the general cargo breakbulk ships. The ocean containers would be loaded and sealed at supply points, and not opened until they reached the buyer's plant docks. The elimination of the multiple handlings of burlap sacks at the docks was expected to bring tremendous cost savings to the coffee bean logistical system. The savings that containerization would bring would also be possible without changing the traditional packaging method.

Test shipments of burlap bags of beans in standard ISO containers began about 1974. By that time the cranes and other facilities necessary to handle ocean containers had been installed at nearly all the major ports. The 20 ft ISO standard ocean container appeared to be ideal for coffee beans. Each container could be loaded with 250–300 sacks of beans to achieve good payloads within the road weight limits of 20 tons, which applies to land transport in most countries.

Ocean container shipments of coffee beans, however, were delayed for several years due to concerns over moisture control in transit. Supply sources for the beans are in tropical zones, and the sacks of beans are loaded in hot, humid conditions. As a ship travels into colder climate zones, the hot moist air from the cargo rises to the ceiling of the container and, under certain temperatures and humidity conditions, will condense and drip onto the sacks of beans. The burlap sacks absorb and hold this moisture, making the beans vulnerable to mold growth and spoilage.

Hundreds of container shipments of green coffee beans were made before a solution to the condensation problem was found. Small nonreversible vent holes, strategically positioned in the container walls, proved sufficient to exhaust most of the hot moist air in transit. The use of absorbent kraft paper over the tops of the loads further protected the cargo from moisture damage during severe climate conditions.

By 1980 ocean container shipments of coffee beans had been implemented for a substantial part of the volume of beans shipped annually. House-to-house movement of the intermodal containers, from supply sources to roasting plants, eliminated multiple handlings and resulted in millions of dollars saved by the coffee bean importers.

The logical next step to improve the logistics for green coffee bean imports would be the conversion of the burlap sacks to packaging more adaptable to mechanical handling methods. The technology of bulk fabric containers had continued to make progress by 1980. By then, many companies in the United States had installed bulk fabric bag systems for many different kinds of granular materials such as sugar, citric acid, food powders, beans, and grains. General Foods had installed shipping and receiving systems at several plants to accommodate interplant shipments of decaffeinated coffee beans.

There was little question that the bulk bag method was cost effective. Handling costs were substantially reduced in comparison with the manual handling of smaller bags. The large strap-handled bags eliminated the need for pallets in the system. Tare weight of the packaging was less than that of an equivalent number of burlap sacks.

The design and size of bulk bags for the import of green coffee beans would have to be worked out. The installation of equipment to fill bulk bags at the sources of supply, and at the receiving locations, would likewise require time and resources to study and implement.

The main objection to proceeding with the development of a bulk bag system for coffee bean imports was that a noncontainerized bulk system should be pursued as a first step. The concept seemed simple enough. Just pump about 20 tons of beans into standard 20 ft ISO containers at the supply source and ship them directly to plants for removal by gravity discharge or airveyor systems. The full bulk system would thereby eliminate the need for, and the costs of, the bulk bags. The logistical system would essentially become automated throughout.

In 1984, in cooperation with two carriers, Sealand and Delta Lines, the Maxwell House division of General Foods carried out a series of feasibility shipping tests for bulk beans in ocean containers. Different kinds of ocean containers were used, as well as alternative methods for preparing the containers for bulk loading and for the discharge systems of unloading. Four countries were involved in the series of tests — Thailand, Brazil, Colombia, and the United States.

Packaging expertise was needed in the preparation of the interior of the ocean containers for bulk loading. The walls and ceiling of the container were lined with a special material made up of starch graft polymer paint on a nonwoven cloth base. This material would absorb excess moisture from the load and prevent damage to the product in transit. Rigid bulkheads, some made of wood and some of heavy-duty corrugated paperboard, were fabricated for the doorways of the containers. This was necessary in order to restrain the loose load. The purpose of the test series was to establish the feasibility of shipping full bulk loads of beans in ocean containers. The crudest loading methods had to be used in lieu of mechanized systems envisioned for later installation if the method was to be successful. Workers manually carried bags into the container, cut them open, and dumped the contents behind the bulkhead. When approximately 36,000 lb (16,329 kg) were loaded, the bulkheads were sealed, and the doors closed and locked. A total of 16 container loads were shipped over a 2-month period to two Maxwell House plant locations. One shipment, arriving at the Jacksonville plant, was unloaded by means of a vacuum system. The other shipments were unloaded by transferring the containers to a tilt chassis, and tilting them to allow the beans to flow out trap doors in the bulkhead to below-ground receiving hoppers.

Although the beans in all the test containers arrived in good condition, the conclusions reached as a result of the tests did not support conversion to the full bulk system for several reasons:

> The cost of preparing the containers, including the bulkhead and lining the walls and ceiling with moisture-control paper, would cost approximately the same as 20 bulk bags.

> Removal of the beans by vacuum was paced by the capacity of the vacuum line to the plant system, which required several

hours to unload a container load. Tilting the container to allow the beans to flow out by gravity was faster, but required a two-person crew and approximately 90 minutes for each container.

Containers that did not have steel corrugated walls bulged under the load pressure. Containers used for bulk shipments would have to be carefully selected and controlled.

Substantial capital investments in filling and unloading facilities, as well as in equipment, would be required to implement a cost-effective bulk system. It was questionable if a reasonable payback on these investments could be obtained on the basis of labor savings from the elimination of the hand dumping of conventional sacks.

A number of administrative issues surfaced that further discouraged going ahead with a noncontainerized bulk program. These included the insurance coverage of bulk loads, the handling of rejected loads for quality reasons, the verification of net weights, and the demurrage on containers when plant silos were filled to capacity. A management decision was made to hold any further plans for the bulk systems development for coffee beans in abeyance.

In 1985, the Gevalia Coffee subsidiary of General Foods in Sweden decided to proceed with a test to determine the practicality and costs to ship green coffee beans in bulk bags. The tests were carried out in cooperation with Nordst Hydro of Porsgrunn, Norway. The firm had developed a bulk bag for the packing and shipping of their industrial agriculture products throughout Northern Europe. The method was so successful that they decided to market it under the trade name, Portabag (Figure 17.1).

The purpose of the test was to determine the feasibility of shipping bulk bags in an ocean container and to establish the size of bag needed, as well as the quantity of beans it could hold.

The bag was specially designed for the test. The details on the bag type and the fill and discharge equipment and procedures are as follows:

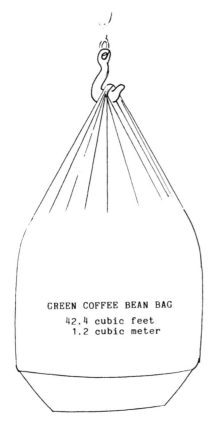

Maximum load — 2240 lbs (1016 kg)
Diameter loaded -- 45 in. (114 cm)
Suspended height -- 72 in. (183 cm)
Load height -- 44 in. (112 cm)

Figure 17.1

Type of bag: The bags were single-handle, top-loop style, constructed of woven polypropylene slit-stretch fibers, UV stabilized. The bags were dimensioned 35 in. (889 mm) long and 18 in. (457 mm) in diameter. There was a filling tube extension at the top and a reinforced double cloth bottom. A safety factor of five to one was used for loads up to a ton.

Filling system: The filling system consisted of a small metal hopper with a bottom spout. It was designed to be carried on the forks of a lift truck. The bag straps were fastened to the base of the hopper, and the filling tube was fitted over the bottom spout. The loose beans were to be conveyed on a feeder belt to the open hopper top. Fully mechanized filling systems would be installed as shipping volume grew (Figure 17.2).

Discharge method: A special hook device was designed to fit over the forks of a lift truck. The top handle loop of the bag was placed on this hook, and the entire bag was lifted and positioned over a receiving hopper. A pyramidal cutting knife was mounted inside the receiving hopper. As the lift-truck operator lowered the bag into the hopper, the knife would cut it open at the bottom to allow the beans to flow out by gravity. The cutting knife had to be mounted well down the inside of the receiving hopper to minimize dust and spillage.

While the filling could have been done at the source of supply in South America, this would have delayed the test many weeks and increase the test costs. A decision was made to divert an incoming 20 ft containerload of coffee beans to the port of Göteborg on the West Coast of Sweden, then unload the sacks and transfer the beans into bulk bags. The bulk bags would then be loaded into the same container and transported overland to the Gevalia plant at Gävle on the East coast.

The filling operations at Göteborg demonstrated the simplicity of the system. The conventional sacks were hoisted to the top of

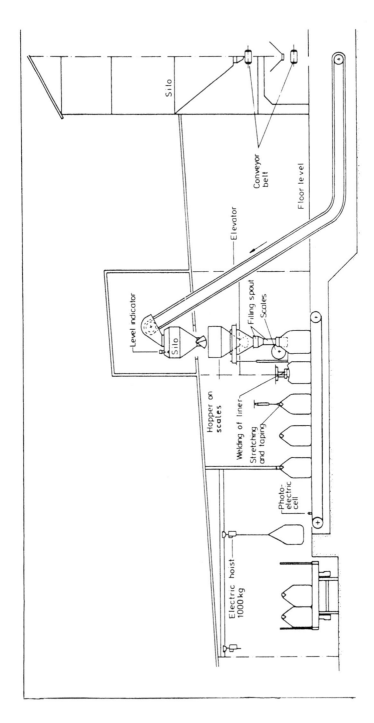

Figure 17.2 Bulk Bag Filing Station. Drawing courtesy Portabulk Handling Systems, N.Y. (Norsk Hydro). Complete filling line installed at Forest Fertilizer, Rjukan, Norway.

218

the fill hopper, cut open, and the beans allowed to fall by gravity through the hopper and into the bulk bags. A conveyorized feed system was envisioned for future filling stations at supply sources.

The bulk bags were each filled with 2002 lbs (908 kg) of beans, which is the equivalent of thirteen 70 kg (154 lb) sacks. The filled bulk bags were stacked two high, and the stacks were loaded into the 20 ft ocean container in a configuration of two across and 10 deep, for a total of 20 bags or 40,040 lb (18,162 kg). Certain techniques for the lift-truck loading process had to be worked out, but the feasibility of one person with a lift truck loading the entire shipment in less than an hour was demonstrated. The hook attachment was sent along with the shipment to Gävle and installed on the forks of a lift truck there. Unloading also required experimentation with techniques, since there was not enough space inside the container to lift the top bags upward by their straps. The technique used to remove the bags on top involved hooking the top handle loop with the attachment hook and tugging it backward until the bag cleared the bottom bag in the stack.

The bulk bag method was proven to be feasible, and a number of potential benefits for it were indicated. These included:

The shrinkage or loss of beans during handling that is common to burlap bags, due to holes and tears caused by the use of stevedore manual hooks, is eliminated with bulk bags. The strong plastic material resists tearing and ripping.

The discharge of beans from the plastic bulk bags is fast and complete. When burlap bags are manually dumped, a few beans cling to the linings and folds.

The plastic fabric material of the bulk bags was found to be sufficiently breathable to allow warm, moist vapors to escape. Condensation, however, could not penetrate the material and cause mold growth.

Tare weight of the plastic bulk bags was approximately 180 lb (82 kg), less than the equivalent number of burlap sacks in the container load.

The sampling of beans for quality assurance examination was done with a small thief sampler device that is pushed through the bag. When withdrawn, the resilient plastic material closes the small hole left under the tension of inside load pressure and prevents the leakage of beans.

Overall system costs for the bulk bag method appeared to be considerably less than either the traditional burlap bag methods or the full bulk alternative. The resources necessary to convert the existing methods to the bulk bag filling and discharge systems appeared to be minimal. The main problem standing in the way of implementing this project was to find someone or some organization to sponsor it and to assume the leadership needed to initiate and coordinate the many changes needed all along the logistical routes. The packaging function in projects of this kind, while critical to the success of the project, has traditionally been a support role. Leadership to implement such projects is usually left to some other functional area. Consequently, indecision and delays continued to frustrate and delay the implementation of bulk plastic bags for imports of coffee beans.

This case history should indicate to packaging professionals and logistics managers that there exists a need to expand the role of packaging for integrated bulk packaging systems. It should not stop with the design and testing of the packaging but should include the leadership necessary to implement the projects.

18

Case History 8: Bulk Containers for Liquid

During the heyday of whaling ships in the nineteenth century, giant wooden casks were used to carry water, juices, and other food products for the crew, and to store and transport the whale oil harvested at sea. Casks consist of a series of wooden staves that are heat warped to create a center bilge. The staves are bound tightly together by a series of metal hoops. The tops and bottoms of the casks are called headings. Craftsmen that make casks are called coopers, and their place of business and products are known as cooperages.

The largest casks aboard whaling ships were dimensioned 30–40 in. (762–1016 mm) across their headings and were 48–50 in. (1219–1270 mm) tall. They contained 175–300 gal (662–1136 l) of liquid, which indicated a gross weight of 1300–2250 lbs (590–1020 kg) for each cask. To conserve space in the holds, the empty casks were dismantled stave by stave and formed into bundles called shooks. A standard shook contained the staves of two large casks and took up less than half the space of two casks set up. On some ships, coopers travelled as members of the crews in order to set up the casks as needed. The staves, hoops, and headings of the casks were identified by numbers to simplify assembly at sea.

Wooden casks provided a marketable product for delivery to ports in the countries where wood was not readily available for their manufacture. Sometimes a whaling ship would deliver a part of its cargo of casks to a buyer in a distant port and keep the rest for the storage of whale oil that was harvested on the return journey.

The loading and unloading of casks was relatively slow. They could be rolled on their sides along the wharfs and hoisted by block-and-tackle gear to and from the ships' cargo holds. The giant casks were transport modules that integrated effectively with the materials-handling methods of the whaling era.

Most containerized bulk liquids today are shipped in rigid metal or plastic drums. The 55 gal (208 l) drum is a widely used standard. Filled drums may weigh from 400 to 500 lb (181–227 kg) depending upon the particular liquid's density. Mechanical handling devices attached to lift trucks are used to grip and lift a cluster of up to four drums at a time. The drums can, therefore, be handled in multiples without the need for pallet bases.

The metal drum containers have certain disadvantages as a transport module for long-haul shipping. They cannot be collapsed when empty and, consequently, are costly to return for reuse. When empty after a long haul, some are sold to local reconditioning companies, where they are cleaned and repaired as necessary for reuse. Others are sold for scrap metal.

The shape and the tare weight of metal drums are other negative features. In comparison with square or rectangular containers, the utilization of storage and cargo space in transport vehicles is poor, and the tare weight of the drums can limit the net payloads over the highway.

The early 1980s saw a flurry of experimentation with bulk transport modules that had the capacity of four to six 55 gal

BULK CONTAINERS FOR LIQUID

standard metal drums. A notable application was the development of an aseptic packaging system for tomato paste in which the bulk container is lined with multiwall plastic liner bags with special valve fittings on the top. The special bags were sterilized by gamma radiation, and an aseptic packaging machine developed by Franrica of California was used to fill the bag with hot liquid tomato paste under sterile conditions.

The container was made of wood with a built-in pallet base and plywood sides, bottom and top. It was sized to contain 300 gal (1136 l) of paste. The outside dimension were $48 \times 44 \times 42.5$ in. ($1219 \times 1118 \times 1080$ mm), and the inside dimensions were $45.75 \times 41.75 \times 38$ in. ($1162 \times 1061 \times 965$ mm). The wood material offered certain advantages for the bulk tomato paste pack. It was strong enough to resist the tremendous lateral pressure that built up inside as the hot tomato paste was injected into the bag. The wood containers could be stacked several high in the fields following filling. The 55 gal metal drums were likewise placed on pallets and stacked outdoors after filling. The outside storage conditions were necessary due to the enormous volumes packed in a harvest season that lasted only 6–8 weeks. On the receiving ends, the inside liner bags were discarded when empty, and the wooden boxes were broken down and assembled into more compact bundles for shipment back to the supply sources. The discharge of the tomato paste at the user locations required the installation of mechanical devices to invert the boxes to allow the paste to flow out by gravity. Some users preferred to use pumping systems. In those cases, the container was tilted to one corner and a wand inserted to transfer the paste out by a vacuum pump.

The first Franrica aseptic bulk container systems were put into operation at three California tomato paste packing plants in 1982. The filled boxes contained the equivalent contents of 5.45–55 gal standard metal drums. The tare weight of the bulk container was approximately 259 lbs (117.5 kg), which was several pounds heavier than the tare weight of 5.45 metal drums. The economic advantage of the bulk wooden boxes over metal drums at that time was a savings of approximately $20 a ton in packaging costs. Additional packaging savings were possible if the boxes were broken down and the parts returned to the supplier for reuse a second time. The liner bags, of course, were usable one time only.

The success of the slipsheeted K box for dry materials stimulated interest in the development of a bulk container for liquids. In 1983 a research and development project was carried out by Containair Systems of New York to determine the feasibility of adapting the paperboard K box to the transport of aseptic packed tomato paste. A prototype K box was constructed with a double-wall corrugated outer box to which a heavy-duty solid fiber slipsheet was bonded. The inner cells were constructed of double-thickness, triple-wall corrugated material. A liner bag was placed in the box and filled with 220 gal of water. The prototype container was then taken to a packaging laboratory and put through a series of dynamic load tests, including incline-impact, vibration, and top-to-bottom compression tests. The lab tests indicated that the adaptation of corrugated packaging to liquid transport was feasible.

Several advantages of the paperboard box were indicated in comparison with either metal drums or the wooden box system. The tare weight would be substantially reduced, and the K Box was expected to cost considerably less per ton shipped. The set up and collapse of the empty containers could be done by one person in a fraction of the time required to erect and later break down the empty wooden boxes. Depending upon the care in handling and shipping, it was expected that the empty used K boxes could be returned and reused a second time.

The potential for the lightweight palletless bulk container for aseptic packed tomato paste led to demonstration shipments in 1984 and 1985. A Chicago manufacturer of barbecue sauce participated in the test program. The production method for the barbecue sauce was based on a batching method that used four 55 gal metal drums for each batch. Consequently, a decision was made to limit the bulk container pack to 220 gal. The length and width dimensions of the corrugated K Boxes were sized 48×44 in. (1219×1118 mm) in order to fit onto the Franrica aseptic filling machine that had been designed for the bulk wooden containers. The height was limited to just 30 in. (762 mm) to provide the cubic capacity to provide the 220 gal (833 l) batch size.

Two prototypes only were filled in the 1983 harvest season. They adapted well to the aseptic filling machine process. In anticipation that the pliable paperboard walls would bulge severely or burst under the aseptic filling pressures, a wooden housing was made to fit around the paperboard box during the filling process. This method, first used to contain lateral pressures during the vibration filling of coffee powder in Brazil (see Chapter 12), worked well and the K Boxes were filled without incident. For comparison purposes, two wooden boxes were filled with 220 gal of the paste and a truckload shipment was made up of the four bulk containers, along with sixty 55 gal metal drums. The shipment was made in a piggyback trailer by rail across the country to the receiving warehouse in Chicago, Ill. The comparative shipping weights of each of the three types of containers was as follows:

	Gross Weight	Net Weight	Tare Weight
220 gal wooden box container	2414 lb	2155 lb	259 lb
	1095 kg	978 kg	117 kg
220 gal K Box with slipsheet	2191 lb	2122 lb	69 lb
	994 kg	963 kg	31 kg
55 gal metal drum	584 lb	539 lb	45 lb
	265 kg	245 kg	20 kg

The four bulk containers were stacked two high in the warehouse and held in storage for a year before opening to satisfy quality assurance requirements. In 1984, the tests were expanded to 200 bulk K Box packs and were shipped in trailerload quantities. All of the boxes performed well, and the feasibility of adapting paperboard boxes to the transport of aseptic-packed bulk tomato paste was generally accepted.

The wooden boxes, however, had one obvious advantage over the paperboard boxes. They could be stored out in the fields and not be subject to collapsing if rained upon. The paperboard boxes would, on the other hand, require warehousing after filling. One possible alternative to tieing up substantial warehouse space for the paperboard boxes was to store the tomato paste in aseptic silos during the harvest season and fill the paperboard bulk containers for shipment to market locations when needed throughout the rest of the year. That, however, would have required a costly start up and a short duration run of the aseptic fillers each time a shipment would be made. Silos existed at the larger packing plants but were fully utilized to store tomato paste for bulk shipments in 22,000 gal bulk tanker cars to very large customers. The conversion to paperboard containers would, therefore, require the installation of additional silos, and the investment in them would have to be justified by the savings obtainable from the use of the paperboard containers in the place of metal drums or wooden boxes.

The difficulties and expected costs of integrating paperboard bulk containers into the tomato paste production and shipping systems of the time discouraged further experimentation with them in 1985. However, the 1983 and 1984 tests demonstrated the feasibility of shipping 220 gal (833 l) of tomato paste in relatively lightweight paperboard containers on paperboard slipsheets. The concept still has considerable potential and is likely to be reintroduced sometime in the future. In the meantime, Containair Systems Corporation has continued to develop and improve their bulk liquid containers for other kinds of products.

Noncontainerized full bulk liquid transport requires the use of dedicated tanker rail cars or trucks. The distances the tankers travel are critical, since in most cases they must return without a load to the shipping source, and the cost of the nonproductive return trip becomes part of the total costs of transportation. Full bulk applications are, therefore, limited to very high volume users of the product.

The quality of the liner bags and the plastic fittings are critical to the performance of bulk containers for liquid (see Figure 18.1). The commercial availability of good liner bags and fittings that prevent leakage appears to be one of the reasons why applications to date have been limited. The production of leakproof liner

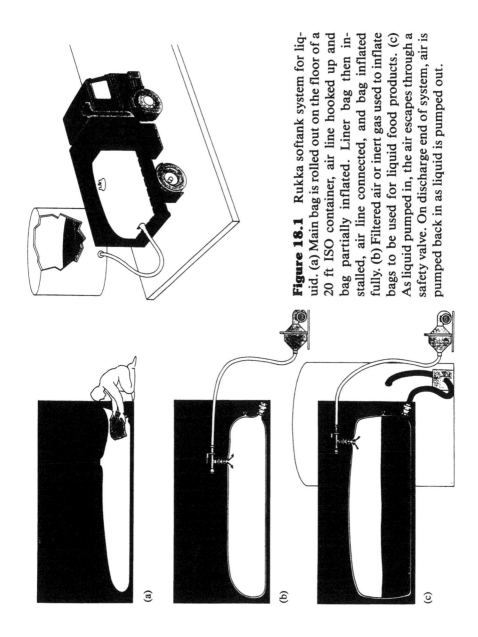

Figure 18.1 Rukka softank system for liquid. (a) Main bag is rolled out on the floor of a 20 ft ISO container, air line hooked up and bag partially inflated. Liner bag then installed, air line connected, and bag inflated fully. (b) Filtered air or inert gas used to inflate bags to be used for liquid food products. (c) As liquid pumped in, the air escapes through a safety valve. On discharge end of system, air is pumped back in as liquid is pumped out.

bags that are adaptable to the 220–300 gal size bulk containers is limited to just a few manufacturers today.

During the 1980s another unique development in container-ized bulk liquid transport took place. It involved the use of plastic fabric containers for the liquid transport of 3170–4226 gal (12,000–16,000 l). Each container holds the equivalent of 58–77 standard 55 gal (208 l) drums, which are truckload quantities. Such capacities lack the flexibility of the smaller bulk transport modules. They require holding and receiving tanks, as well as special pumping facilities at the shipping and receiving locations. They offer an alternative to truckload or ocean containerload vol-umes of standard drums, and they provide greater logistical flexi-bility than dedicated bulk tanker systems. Certain elements of their design may be applicable to the design of smaller bulk mod-ules in the future.

The story of the development of one of these systems, the Rukka bag, provides an especially interesting case history of inte-grated bulk packaging system design. The Rukka bag, which was under development for several years, is now commercially availa-ble for international transport of nonhazardous liquids in 20 ft ocean containers.

The developer had no particular application in mind for it when the project was initiated. It was developed to satisfy a need for a low-cost bulk container system to transport liquids in dry cargo vehicles. The key economic factor was the potential for in-creased utilization of the vehicles to produce greater revenue, along with the capability to carry either dry or liquid cargo.

The idea came out of a brainstorming session in 1979 at Rukka Oy, a manufacturer of air-supported buildings and water-proof clothing in Kokkola, Finland. The objective of the session was to get ideas for new products that would employ the use of the company's design skills and expertise in the manufacture of inflat-able waterproof fabrics. The idea seemed simple enough at the time. It was to inflate a large fabric bag with filtered air and then pump in liquid to displace the air. The liquid-filled "balloon" would then be shipped to a distant location, where air would be pumped back in as the liquid is pumped out. When empty, the bag would be deflated and folded into a small package for use for another trip.

The idea was not entirely new since flexible containers made of rubber and canvas had been used for liquids for some time. What was new was the idea of keeping the bag inflated during the pumping and the discharge of the product. This was a key feature that could overcome the problems inherent in pumping liquid in or out of a collapsing flexible bag.

As packaging engineers began to research the project, they learned of a key disadvantage to the use of such containers. The initial investment in the packaging was relatively high, and, consequently, the bags had to be reused several times to be affordable. If used for food products such as fruit juices, they have to be cleaned and sterilized before reuse. That, along with the weight of very large bags, added to the recovery and return costs, and reduced the revenue they could produce. Rukka packaging engineers pursued the development of a lightweight double-bag system to keep the recovery and return costs of the containers to a minimum.

That led to the idea of using a container made up of two bags. There would be a tough outer bag of a durable plastic fabric and an inner disposable liner bag of less expensive plastic material. The liner bag would be disposed of after each trip, thereby eliminating the time and cost of washing and sterilizing the main bag. The outer bag with its fittings would be light and flexible enough to fold up into a compact package. That would make it possible to ship it back to the supply source by air freight and would minimize the total number of bags required in an ongoing system.

When Rukka made known its development effort in 1981, it caught the attention of engineers who had been working on a similar project at the Johnson Line in Sweden. Unknown to Rukka at the time, Johnson had been working with British Hovercraft on a similar system since 1980. Their incentive was to find a means of utilizing empty ocean containers for return trips from Brazil to Europe. Frozen orange juice concentrate, known in the trade as frocon, was of particular interest. At the time that frocon was being shipped in 55 gal (208 l) polyethylene-lined steel drums in the refrigerated holds of breakbulk ships. Due to the huge volumes shipped, an alternative under development at the time involved the use of bulk tanker vessels.

Johnson reasoned that an effective low-cost packaging system, used in conjunction with intermodal containers, could compete favorably with bulk tankers, since the intermediate holding tank and transfer facilities that would be required at ports could be eliminated. They believed that would not only reduce total system costs substantially, but also would improve product quality, since the product would be put through only two pumping operations in the system.

The intermodal containers would be consigned directly from the supplier plants in Brazil to the users in Europe. The elimination of empty return trips of the dedicated bulk vessels would further reduce the overall system costs per liter shipped.

Like Rukka, Johnson planned to use a double bag, a cheap inner liner and an outer bag made of durable, Hovercraft skirt material that would be reused many times. Shipping tests in the fall of 1980 tested the double-bag concept, and a number of technical problems surfaced. A prototype filled with frocon in Brazil resulted in inner-bag rips and tears, as well as leakage between the inner and outer bag. During the discharge pumping operation, the collapsing bags formed large pockets that trapped the product and rendered much of it nonpumpable. It was then that it was realized that the inflatable double-bag concept of Rukka was needed.

Johnson dropped the Hovercraft experimental project, and they joined with Rukka to pursue the development of the inflatable bag concept. The shipment of frocon in ocean containers provided a specific application incentive, consequently, the system that evolved was designed to be compatible with the standard 20 ft ISO intermodal ocean container (Figure 18.2). This resulted in the sizing of containers to carry 12,000 l (3170 gal) in reefer containers and 16,000 l (4227 gal) in dry cargo containers. These capacities would provide for excellent utilization of the containers and for the economical payloads of liquid products within the over-road weight limits for land transport of the intermodal containers.

Figure 18.2   Rukka softank system in container.

The design of the bags and fittings was modified as necessary to precisely fit the ocean containers so they would not shift about during transport but, instead, would remain securely in place without a need for lashing.

PVC-coated polyester was selected for the outer reusable bag, with all seams double constructed and high-frequency welded. The inner liner bags were improved and were made of polyethylene film, which was treated with gamma radiation and approved for food products.

Prototype bags and fittings were manufactured for a series of shipping tests over a 2-year period of time. Included in the test shipments were fruit juices from Europe to New York, fish oils from Iceland to Great Britain, frocon from Florida to New York, and USP glycerine from Colombia to California.

In 1987 a prototype bag was sent to the Helsinki Sörnäinen railway testing yards for a series of rail transit tests. The bag was installed in a 20 ft ISO container and placed onto a rail flatcar. Altogether 15 collision impact tests were made in a speed range of 1.5 to 7.5 mph. The required 2G stress level inside the container was exceeded seven times. The bag withstood all impacts and was awarded a certificate of approval for rail transport in Finland.

Improvements in fittings and bag design continued to be made during the 2 years of shipping tests. By 1987 the system was considered to be commercially marketable. It was given the trade name, Rukka 2 in 1 Softank System. Axel Johnson Marine of Helsinki became the marketing agent for the worldwide distribution of the bags.

The Rukka Softanks are delivered to users in a compact plastic-fabric covered Styrofoam envelope measuring $33 \times 24 \times 20$ in. $(838 \times 610 \times 508$ mm). The package weighs 157 lb (71.2 kg) for the 12,000 l size and 200 lb (90.7 kg) for the 16,000 l size. It contains all the necessary parts, including the bag and disposable inner bag liners, the fill discharge valves, air supply fittings, a hatch sealing strip, and the necessary tools.

The first step in the set-up procedure is to roll the outer bag out onto the floor of the ocean container and partially inflate it. The inner liner bag is then inserted through a hatch opening at the rear of the outer bag and secured with its fittings and valves. The hatch is then clamped shut, and the liner is inflated with filtered air. When fully inflated, the liquid product is pumped in to displace the air. In this way, the unit maintains its shape through the filling process. When the bag is filled, the valves are closed, the hoses are disconnected, and the load is ready for shipment.

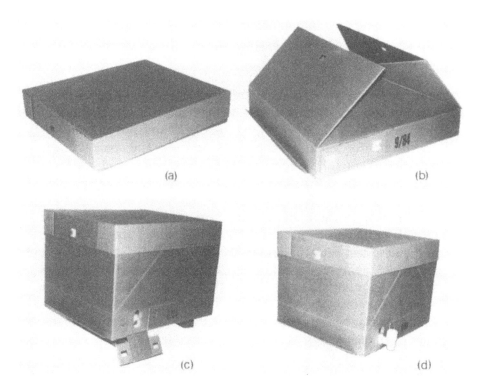

Figure 18.3 KL box – corrugated box with slipsheet designed for liquid transport – 220-300 gal (833-1136 l) multi wall plastic bag with fitments inside. Photo courtesy Containair Systems Inc., New York.

Upon arrival at the receiving plant, the liquid is pumped directly to receiving tanks, where filtered air is again pumped into the inner bag to keep it from collapsing as the liquid level depletes. When discharge is completed, the inner liner bag is pulled out and discarded. The outer bag and fittings are packed into a compact transport envelope for return to the supply source for immediate reuse. The size and weight of the empty return pack make it suitable for return by air freight to speed its recovery and reuse. The manufacturer states that users may expect 25 or more round trips from the outer bag and an indefinite number of uses of the metal fittings. This, of course, would depend upon the care given in handling.

Although the system was designed for intermodal ocean container transport, its first commercial application was for an overland trucking operation between Finland and Norway. A trucking company, Ahola Transport Oy in Kokkola, found in it an opportunity to improve the utilization of its vehicles. The compact bags were carried aboard the company's dry cargo trucks and trailers in order to convert them rapidly to tankers that could carry liquid cargo, such as fish oils or emulsions, along the route.

In 1988, research and development for improvements to the Softank containers continued. Most of the work today is centered on the development of improved and less costly fittings and inner liner bags. Lightweight containers for the shipment of bulk liquids are needed for both the 220–300 gal (833–1136 l) size, as well as the larger bulk tank-size types. More widespread use of the concepts will depend upon improved technology and reduced costs for the inner liner bags as well as the outside containers.

The case history of the Rukka bag and the tomato paste containerized bulk transport module for liquid transport illustrates the amount of research, time, and expenditure that is necessary to design packaging that can be effectively integrated to suit the needs of the total logistical system.

19

Case History 9: Unitized System Development

Most major projects that involve packaging, shipping, and handling are initiated and introduced to higher management by any one of a number of functional areas in a large company. Some projects may originate in the logistics or physical distribution departments. Others may come through industrial engineering, operations management, marketing, purchasing, or other staff areas. The packaging department generally provides technical support.

If the role of packaging is expanded to include more direct involvement in the initiation and implementation of major logistical projects, then additional responsibilities go along with that role. Included will be not only a good deal of research and study of the total system, but the communication of economic opportunities that will put top management in a position to make decisions on resources and support for the program.

The case history of Apple Computer's SPUDS program (Supporting Palletless Unitized Distribution System) illustrates what is involved in initiating and carrying out a major project that includes primary packaging modifications and the design of palletless transport modules. The industrial engineering area, called the Distribution Engineering Department at Apple, provided the project leadership to initiate and implement the total program.

The project had its beginnings in a series of problems that were being experienced in the receiving of inbound ocean containers of computer products from foreign suppliers. Early in 1987, the Distribution Engineering Department held meetings to explore what could be done to eliminate the unacceptable expenses and processing delays in manually unloading ocean containers from Far East suppliers.

Most of the deliveries to Apple Distribution Centers were floor loaded, carton by carton, in 40 ft ocean containers. That meant, of course, that all of the cartons that were delivered had to be hand transferred onto pallets at the receiving docks. This process took 5–7 hours to unload each container and tied up valuable space at the docks that was needed for the outbound shipments of customer orders.

Several decisions and conclusions were reached in these initial meetings, which included:

1. The products would have to be received unitized in order to speed the unloading operations.
2. Pallets would be impractical for international shipments of this type due to the reduction in net payloads caused by the space the pallets occupied under the loads. This would result in substantial transportation penalties (see Figure 19.1).
3. The slipsheet unitized system appeared to be the best alternative available. Previous trials of the sheets in 1984, however, indicated problems with slipsheet handling that would have to be overcome.
4. Partial conversion to slipsheets was considered to be impractical. A successful program would depend upon the

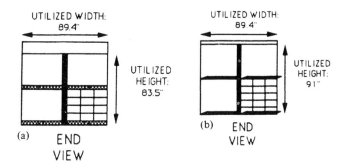

	Pallet	Slipsheet
Module size	48.5×44.7×36.4 in. (1232×1135×925 mm)	44.7×32.2×45.5 in. (1135×818×1156 mm)
Carton size	16.1×14.9×9.1 in. (409×379×231 mm)	16.1×14.9×9.1 in. (409×379×231 mm)
# of modules per container	44	68
# cartons/module	36	30
Total # cartons	1584	2040
Carton weight	15 lbs (6.8 kg)	15 lbs (6.8 kg)
Net weight/container	23,760 lbs (10777 kg)	30,600 lbs (13880 kg)

Figure 19.1   Pallet versus slipsheet comparison.

participation of all vendors and shipping and receiving locations throughout Apple's global network.

5. To succeed, the slipsheet program must have the backing of top management people and their commitment to the resources necessary to progress such a program. The economic reasons for the conversion to slipsheets must, therefore, be thoroughly researched and communicated effectively to management. Industry trends and the impact that slipsheets will have on the other functional areas such as purchasing,

transportation, packaging, warehousing, sales, and marketing would all have to be addressed.

6. Resistance to change to the more sophisticated slipsheet
method could be expected. Therefore, the program would
require a major educational effort to convert foreign suppliers, vendors, freight carriers, and warehousing people to
the method.

The decision to convert to slipsheets directed emphasis to the
material-handling activities throughout the total system. The cost
feasibility would depend in part upon achieving the same or better
load densities with the cartons unitized on slipsheets versus hand
stacking methods. The immediate approach was to adapt the existing carton sizes to unitized patterns and to select slipsheet module
sizes that would provide at least the same number of cartons per 40
ft ocean container as did the hand-loaded method. Figure 19.2
gives a comparative analysis of one of the product types hand
stacked versus unitized on slipsheets. Since it indicated that there
would be no loss in payload, there would be no transportation
penalty applicable to the unitized method.

Subsequent studies indicated that slight changes in the carton
dimensions, along with the improved load patterns that the
changes made possible, could substantially increase the number of
cartons per shipment. Figure 19.3 provides an example of how a
slight change in carton dimensions can impact the total load density. In this example a 0.4 in. (10 mm) reduction in the height of the
cartons made it possible to get an extra tier of cartons on the top
modules in the load. This increased the number of cartons per
shipment by a total of 88, which meant 9% fewer container shipments would be needed for the same volume of this product. The
revision to other carton dimensions, sizes of slipsheets, and unitized patterns resulted in some 250 fewer containers shipped each
year for the same volume.

Apple's total program involved the conversion of a worldwide
system that stretched from the Far East to Europe. Distribution
engineers were sent to visit the overseas vendors and manufacturing sites to research and prepare costs and potential savings analy-

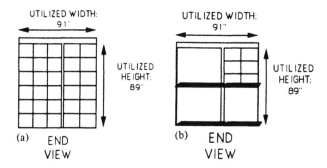

Type of shipping container - ISO 40 ft
Container ID - 474×92×93.7 in. (1204×234×238 cm)
Carton dimensions - 19.4×17.2×14.8 in. (493×437×376 mm)

	Handloaded	Slipsheeted
Slipsheet dimensions	N/A	52×39×44.5 in. (132×99×113 cm)
# cartons/module	N/A	18
Load configuration	126/layer 6 tier high	21 stacks of two modules – 42 total
Total # cartons	756	756
Total net weight per container load	20,412 lbs (9259 kg)	20,412 lbs (9259 kg)

Figure 19.2   Handload versus slipsheet modules.

ses. Presentations were then given to sell the program to the foreign sources, and arrangements were made for the initial supply of slipsheets and for the acquisition of push-pull attachments.

To assist in the installation of equipment and in the training of lift-truck operators, Apple called upon the Cascade Corporation. This company manufactures push-pull attachments that are marketed through lift-truck dealers. Cascade maintains a world-wide network of representatives to assist the start up of their equipment in far-flung locations. As many as 10 people from Cascade were heavily involved in Apple's conversion program.

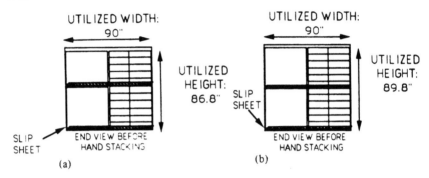

Figure 19.3 Impact of resized carton.

	Original Carton	Resized Carton
Carton dimensions	22.5×19.8×[7.8 in.] (572×503×198 mm)	22.5×19.8×[7.4 in.] (572×503×188 mm)
Module size	45×40×39.5 in. (1143×1016×1003 mm)	45×40×44.4 in. (1143×1016×1128 mm)
Configuration	4 cartons/tier 6 tier on bottom modules and 5 tier on top modules	4 cartons/tier 6 tier—all modules
# cartons/module	22 average	24 all
# modules/40 ft container	44	44
Total cartons/container load	968	1056
Carton weight	17 lbs (7.7 kg)	17 lbs (7.7 kg)
Net shipment weight	16,456 lbs (7464 kg)	17952 lbs (8143 kg)

With the installation of proper equipment, and the training of lift-truck operators, 40 ft containers could be loaded or unloaded by one man in just 30 minutes (Figure 19.4). This compared with an average of 6 man hours per container to load or unload by hand.

While the primary interest in slipsheets was for shipping, the slipsheet modules could be stacked in warehouse storage areas the

Figure 19.4 Unloading modules from an ocean container. Photo by
Lyle Crouse, courtesy of Apple Computer, Cupertino, Cal.

same as palletloads (Figure 19.5). The conversion to slipsheets,
thereby, eliminated the use of pallets for storage and for the transport of modules to container loading docks. This improved the
utilization of warehouse space and added further to the program's
total savings. There was also a noticeable reduction in product
soilage and damage with the slipsheeted modules.

The major economic benefits of the slipsheet unitized shipping program included a 20% freight savings due to the higher
density loads, a 75% reduction in labor costs to unload containers,
and a 78% reduction in pallet costs. The total savings for Apple
over the first year the program was installed amounted to over $2
million, and when fully installed the program realized over $3
million annual savings. By 1989, 94% of the company's products
arrived at the distribution centers on slipsheets. Apple's manufac-

Figure 19.5 Storing Modules in Apple Computer warehouse. Photo by Lyle Crouse, courtesy of Apple Computer, Cupertino, Cal.

turing plants quickly became part of the program and also started to receive slipsheet loads of subassemblies and components.

The success of the internal shipping program inspired Apple to offer deliveries to their customers on slipsheets. It was found that many of the large customers were reluctant to purchase and install push-pull attachments on their lift trucks to receive the products of a single computer company. Intent on taking the program all the way to its customers, Apple then decided to publicly make known the results of its slipsheet program in order to interest other computer product companies in the method. It was felt that as other companies joined the program, Apple's customers would be more willing to accept deliveries on slipsheets. Apple subsequently held an open symposium for the computer industry in 1989 for the purpose of sharing their experience with other companies.

During the symposium, Apple people outlined a list of steps that were considered necessary to implement a successful slipsheet module shipping program. Included were:

1. Assignment of a functional area to provide management and leadership for the entire program. At Apple this was the Distribution Engineering Department.
2. Top management commitment to the program.
3. Orientation programs for all people who would be impacted by the change from existing methods throughout the logistical system.
4. Ongoing communication to sales and marketing departments. These areas are specially interested in any program that can impact customer relations. With complete understanding of the program objectives, the sales and marketing people can contribute a supportive role.
5. Planning schedules for the acquisition of equipment and slipsheets, installation, training of operators, and start up are essential. Implementation should be staged with specific schedules and benchmarks.
6. Follow-up at each stage to correct any problems that could undermine enthusiasm for the program was found to be critical. Likewise, periodic reporting of the savings achieved as the implementation progressed motivated those participants who were finding the conversion difficult.
7. Keeping communication open and flowing between all members of the team was helpful and resulted in maintaining the spirit of teamwork and a continuing interest in the program.

Packaging, transportation, and other functional areas all played supporting roles in the development and implementation of the Apple program. Since the design of the modules and the resizing of cartons were most important to the total program, the question may be asked whether or not packaging people should play a more dynamic role in projects of this kind. Integrated packaging design is critical to the success of unitized shipping. Top management will look to the functional area that they believe will get the job done. Therefore, packaging people must be equipped to go beyond the packaging lab if they are to be recognized by top management as having the capability to handle the development of total logistical systems.

Information in figures compiled from materials presented at Apple Computer's Material Handling Symposium, March 1, 1989 at Santa Clara, California. The materials included a paper by Patricia Lyons and Dale Spenner titled: "Slipsheets Generate Hyper-Savings at Apple Computer."

20

Summary

When the first unitized shipping program was implemented in the food industry in the 1950s, physical distribution and logistics departments of large companies were not yet in existence. *Logistics* was a term associated with military operations, including the procurement, storage, and transportation of military supplies, as well as the movement of military equipment and personnel.

The corporate traffic departments of large companies played a key role in the implementation of the first unitized shipping programs, since negotiation with carriers was a major part of the effort. Research and development activities for unitized shipping were usually assigned to the industrial engineering departments of companies that had such specialists at the time. Packaging specialists were concerned mainly with the development of the primary package specifications and had little to do with the design of the transport modules.

In the years that followed, physical distribution and logistics emerged as important new management functions to plan and control the flow of goods and materials in industry. In 1985 the Council of Physical Distribution Management was renamed the Council of Logistics Management, and it adopted the definition of logistics as, "The process of planning, implementing, and controlling the efficient, cost effective flow and storage of raw materials, in-process inventory, finished goods, and related information from point of origin to point of consumption for the purpose of conforming to customer requirements." The definition encompasses just about everything concerned with the inbound, outbound, internal, and external movements of goods and materials. Within this broad scope of activities are a number of independent functional areas, each of which has its own cost centers and management objectives. In most large companies, packaging, materials handling, warehousing and storage, transportation, and distribution are autonomous parts of the total logistical system.

Specialists in each of these areas have different kinds of work experience and academic backgrounds. There is little cross fertilization. Their areas of expertise have their own separate professional societies, trade publications, trade shows, and conferences. As a result, specialists in any one activity may not be fully aware of the needs and goals of those in the other activities in the total system. The mission of logistics management is to bring it all together for the common good of the system.

This book has dealt with the contribution of packaging to the logistical process. It concludes that packaging design is an elemental and crucial part of the logistical process that can impact the efficiencies of all other activities. Packaging specialists are, therefore, in a position to play a broader role in the design and implementation of logistical systems.

In order to take on a broader role, it is important that packaging specialists understand the functional and cost objectives of all of the independent areas of activity in the entire logistical process and design packaging as a system.

Chapters 11 through 19 include case histories of logistical systems in which packaging design played a key role. The projects were introduced by functional areas other than packaging, but packaging specialists provided technical support in each case.

There probably is no one technical specialist area in which the knowledge and expertise exists to design or plan for all the functions in a total logistical system. However, there are certain research guidelines that can help the packaging specialists to design packaging for transport modules that will integrate well with the other tangible activities in the logistical chain. The following steps summarize a systematic approach to the research and other activities necessary for effective design and integration of packaging into a total logistical system:

1. Research all pertinent details of the item or material to be packaged and shipped that will influence the selection of packaging materials. Obtain samples of the product.
2. Research all pertinent details of the logistical process through which the product will move. Include the volumes to be shipped, the frequency of production and shipment, shipping routes, and the modes of transportation. In the process, check out the inside and doorway dimensions of transportation vehicles, types of warehousing and storage facilities, along with the materials-handling methods and equipment that is available at all shipping and receiving locations in the chain.
3. Check for any governmental regulatory requirements for the type of packaging for which compliance is mandatory. Information of this kind may be obtained through the trade associations that represent each transportation mode or through the particular government agencies. (A list of associations and government agencies is included in the bibliography.)
4. Establish the ideal or optimal transport module dimensions and weight. In making this determination, the efficient utilization of transportation equipment and warehouse space must be considered. Determine the options open to the dimensioning of the smaller units in the module to form into compatible load patterns. Sometimes a minor change in a

primary unit dimension, or a rearrangement of a case-pack configuration, will result in an improved transport module.

5. Determine if opportunities exist for improving pack densities in order to reduce the costs of packaging per net ton shipped. There are different ways to increase the density of the load. Chapter 16 provides an example in which two large automotive parts were nested together in order to get as many pieces as possible within a shipment. Chapter 12 describes the use of a unique vibration system that can be used to deaerate powders and granules during filling in order to improve the pack density.

6. Construct prototypes of the transport module, whether it be a unitized load of smaller units or a single, large bulk container. Select a packaging test facility that has equipment large enough to test full-size transport modules. Put the prototypes through a series of compression, shock, and vibration tests that will simulate the expected transit and storage conditions. Give attention to any weakening and bulging of container walls that may take place that could limit the number of modules shipped in a vehicle or result in inefficient utilization of warehouse and storage space. Build improved prototypes as necessary until lab test results are satisfactory.

7. Construct a sufficient number of prototypes for a field test and make arrangements for a test shipment. Monitor all phases of the test shipment and evaluate the performance of the packaging and the transport module throughout the system. Modifications to improve the structure of the packaging or the functional performance of the modules may be required before the system is ready for implementation.

8. The implementation process of major projects should involve packaging people in the orientation and training of operating personnel to ensure that they understand the procedure for the preparation of the transport module. If a bulk container module is involved, instructions for the set up, filling, and closure of the containers should be provided to all operating personnel. Likewise, instructions should be

provided at the receiver's location for the handling, opening, and discharge of the load, as well as the handling of the empty container.

The above guidelines indicate an expanded role for packaging specialists. Packaging and transport modules that are designed to interface efficiently with all other activities in the logistical system will be critical to the improvement of the productivity of human and material resources in the movement of goods and materials in a steadily increasing and more interactive world population.

Glossary

ANSI	American National Standards Institute
Aseptic pack	packaging method in which sterilized product is packed under sterile conditions into sterilized containers.
ASTM	American Society for Testing & Materials
ATA	American Trucking Association
Captive pallet	pallet used only at specific location, i.e., a nontransit pallet.
COFC	Container on flat car; most common are 20 and 40 ft ISO intermodal containers
Demurrage	a penalty paid for holding a transport vehicle a longer time than allowed by the applicable tariff

Dunnage	materials used to fill voids between pieces in a shipment in order to stabilize the load and prevent damage in transit
DFB	*D*amage-*f*ree *b*ulkhead type rail car that is specifically designed for the shipment of transport modules
EEC	European Economic Community – the common market
FIFO	*f*irst *i*n, *f*irst *o*ut, storage method; opposite is LIFO – *l*ast *i*n, *f*irst *o*ut
Fifth wheel	the round horizontal metal plate just over the rear wheels of a highway tractor on which the front end of a semitrailer is mounted.
GMA pallet	Grocery Manufacturers Association standard wooden pallet for the food industry exchange program
IATA	International Air Transport Association
IBC	a generic term for large bulk containers meaning *i*ntermediate *b*ulk *c*ontainer
ICHCA	International Cargo Handling Coordination Association
Intermodal	trailers or large containers that are interchangeable between two or more modes of transport, such as ship, rail, and truck for ISO ocean containers, or highway trailers on flat cars (TOFC)
ISO container	containers with standard specifications that are established by the International Standards Organization; most common are 20 ft and 40 ft containers

IQF   pieces of fruits or vegetables that are quick
                        frozen as individual pieces instead of a solid
                        block

K boxes   series of large, bulk corrugated boxes that are
                        designed to collapse when empty into the
                        base dimensions; slipsheets are permanently
                        mounted on the bases

Pul-Pac   a registered trade name used by Clark Equip-
                        ment Company as a term for the original
                        slipsheet handling method and equipment;
                        the generic term for slipsheet handling
                        attachments today is *push-pull*

RBL   a rail car with bulkheads that are the same as
                        a DFB car but temperature controlled for
                        shipment of frozen products or products that
                        require air conditioning in transit

RSC   corrugated containers that have full flaps top
                        and bottom

TOFC   *trailer on flat car* — known commonly as
                        piggyback in the United States

Tote box   a commonly used trade term for a large bulk
                        box

UV stabilized   plastic material that has been treated to
                        prevent deterioration by exposure to the
                        ultraviolet rays of the sun

WPO   World Packaging Organization

References

Preface

1. General reference: J. H. Kemble, *S. F. Bay*, Cornell Maritime Press, Cambridge, Maryland, 1957.

Chapter 1

1. G. F. Bass, Oldest known shipwreck reveals Bronze Age splendors, *National Geographic*, December (1987).
2. W. F. Friedman, J. J. Kipnees, *Distribution Packaging*, Robert E. Krieger, Huntington, NY, 1977 (general reference material on distribution packaging).

Chapter 2

1. C. W. Ebeling, Instant pallet patterns, *Modern Materials Handling,* *25*, 8, August (1970).
2. Computer selects most economical patterns for pallet shipment, *Management in Practice* (Newsletter of American Management Association) June (1980).
3. *Another Long Step Forward*, promotional literature of General Foods Palletized Shipping Program, 1962.

Chapter 3

1. C. W. Ebeling, Skee sheets: a shipping alternative. *Transportation and Distribution Management, 30*, 12, November (1987).

Chapter 4

1. C. W. Ebeling, Technical paper presented at the *Annual National Packaging Forum*, Chicago, Ill. Nov. 1, 1972.
2. C. W. Ebeling, Creative application of a new technology to an old concept, *Handling & Shipping, 12*, 10, October (1971).
3. C. W. Ebeling, Packaging for unitized shipment, *Australian Packaging*, Feb. (1974).
4. General Foods Peels Warehouse Costs, *Traffic Management*, Feb., 63 (1969).

Chapter 5

1. General reference materials: *American Paperboard Institute*, New York City, N.Y.
2. C. W. Ebeling, Push-pull or pallets? Distribution's Dilemma, *Handling & Shipping Mgmt.*, Oct. (1978).
3. C. W. Ebeling, Why General Foods Converted to Slipsheets, *Grocery Distribution, 3*, 5, May/June (1978).
4. C. W. Ebeling, Push-pulls global progress, *Handling & Shipping Mgmt., 20*, 12, Dec. (1979).
5. *American National Standard ANSI MH1.5M* — 1980 American Society of Mechanical Engineers, New York, NY, 1980.

Chapter 6

1. R. W. Porter and R. Swatton, Intermediate bulk containers, an ICHCA Survey, *ICHCA*, Abford House, London, U.K. 1979.

Chapter 8

1. R. Roberts, The future is intermodal, *Modern Railroads*, Nov. (1979).
2. M. Felice, The dawn of containerization, presented at *Containerization and Intermodal Institute 25th Anniversary Luncheon*, Vista Hotel, New York World Trade Center, May 5, 1981.
3. General reference: *American Bureau of Shipping*, New York, NY. Surveyor, *18*, 3, Nov. (1984).

Chapter 9

1. C. W. Ebeling, Raising the roof on warehousing, *Handling & Shipping Mgmt.*, May (1980).
2. Editorial feature article, Concrete for high racks, *Modern Materials Handling, 29*, 8, August (1974).
3. C. W. Ebeling, A practical design concept for needed innovation in warehousing, *Handling & Shipping, 14*, 2, February (1973).

Chapter 10

1. ASTM Subcommittee D10.22 on Handling & Transportation Designation D 4169. *Mechanical Handling of Unitized Loads and Large Shipping cases and Crates*. See part 11.3 Element B—Mechanical handling over 100 lbs (45.4 kg). Obtainable through the American Society for Testing and Materials, 1916 Race St., Philadelphia, PA 19103.

For information on packaging design for compliance to domestic and international trade rules and regulations contact following:
Air Transport Association of America
1709 New York Ave., NW
Washington, DC 20006

Association of American Railroads
50 F. Street, NW
Washington, DC 20001

American Trucking Association
2200 Mill Road,
Alexandria, VA 22314

UNIDO
P.O. Box 300
Packaging Sector, A-1400
Vienna, Austria
Attention: Director

U.S. Department of Transportation (DOT) Research and Special Programs Administration Code of Federal Regulations Parts 178 to 199 on Packaging. Published by the office of the Federal Register National Archives and Records Administration. For sale through The Superintendent of Documents, U.S. Government Printing Office, Washington, DC 20402.

UNCTAD/GATT International Trade Center
54–56 Rue de Montbrillant
CH-1202 Geneva, Switzerland
Attention: Advisor, Packaging

EEC: Chief Information Specialist, Press and Information
Delegation of European Communities
2100 Elm Street NW, Suite 707
Washington, DC 20037

Chapter 11

1. C. W. Ebeling and K. Robe, Incoming shipments of raw materials in corrugated fiber bins, *Food Engineering*, Nov. (1983).

Chapter 12

1. Editorial feature, General Foods finds new way to ship incoming ingredients, *Packaging*, May (1984).
2. Slipsheet tote boxes revolutionize soluble shipments from Brazil, World, *Coffee & Tea, 25*, 5, September (1984).
3. M. Borsellino, Shipping containers, *Materials Management and Distribution (Canada)*, December, 29, (1984).

Chapter 13

1. Historical information provided by James E. Wampler, VP, Basiloid Products Co. and Ralph McGaughey, formerly General Electric Facilities Planning and Project Engineer, Louisville, Ky.
2. C. W. Ebeling, The Convertible Container, *Handling and Shipping Mgmt.*, *27*, 8, August (1986).
3. Editorial feature, Final touch in packaging, *Transportation and Distribution Mgmt.*, Dec. (1987).
4. C. W. Ebeling, Top lift handling still viable, *Transportation and Distribution Mgmt.* May (1989).

Chapter 14

1. C. W. Ebeling, Gravity buggies make the grade, *Handling & Shipping*, January (1978).

Chapter 15

1. C. W. Ebeling, The convertible container, *Handling & Shipping Mgmt.*, *27*, 8, August (1986).

Chapter 16

1. C. W. Ebeling, Transatlantic: With a stretch of the imagination, *Handling & Shipping Mgmt.*, *28*, 3, March (1987).

Chapter 17

1. C. W. Ebeling, Productivity in the bag, Handling & Shipping Management, March (1981).
2. R. Kato, President, Japan Packaging Institute, Results of survey on the state of use of flexible bulk containers, September (1982).
3. C. W. Ebeling, Changes brewing in coffee industry, *Handling & Shipping Mgmt.*, *27*, 7, July (1986).

Chapter 18

1. C. W. Ebeling, Seagoing balloons, Transportation & Distribution, April (1988).
2. Information on bulk casks researched at Mystic Seaport, Mystic, Conn.

Chapter 19

1. D. W. Yocam, Technology takes bite of success, Transportation & Distribution, presidential issue (1988–89).

Index

For Product Safety Concerns and Information please contact our EU
representative GPSR@taylorandfrancis.com
Taylor & Francis Verlag GmbH, Kaufingerstraße 24, 80331 München, Germany

www.ingramcontent.com/pod-product-compliance
Ingram Content Group UK Ltd.
Pitfield, Milton Keynes, MK11 3LW, UK
UKHW021618240425
457818UK00018B/632